A COLD WAR AND
A HOTBED OF SECRETS

On April 16, 1996, the *New York Times* reported that a mysterious underground complex was being built by the Russians in the Ural Mountains. The project was so big, tens of thousands of workers were involved and an entire highway and new railroad extension had to be built in order to service it.

U.S. intelligence sources believe the Russian government pumped more than $6 billion into the Yamatau mountain complex, a cavern that spans some four hundred square miles. One theory suggests that in the event of a nuclear war the Russian leadership would head for Yamatau, a place where they could survive until the after-effects of such an apocalyptic conflict settled down.

But there might be more to it than that, especially considering, as we will see, other unusual UFO-related events going on inside Russia.

Could Yamatau be another Area 51 located deep inside a mountain? Literally a "Russian S4"?

No one really knows.

BEYOND AREA 51

Mack Maloney

BERKLEY BOOKS, NEW YORK

THE BERKLEY PUBLISHING GROUP
Published by the Penguin Group
Penguin Group (USA) Inc.
375 Hudson Street, New York, New York 10014, USA

USA | Canada | UK | Ireland | Australia | New Zealand | India | South Africa | China

Penguin Books Ltd., Registered Offices: 80 Strand, London WC2R 0RL, England
For more information about the Penguin Group, visit penguin.com.

BEYOND AREA 51

A Berkley Book / published by arrangement with Kelcorp, Inc.

Berkley Books are published by The Berkley Publishing Group.
BERKLEY® is a registered trademark of Penguin Group (USA) Inc.
The "B" design is a trademark of Penguin Group (USA) Inc.

For information, address: The Berkley Publishing Group,
a division of Penguin Group (USA) Inc.,
375 Hudson Street, New York, New York 10014.

ISBN: 978-0-425-26286-3

PUBLISHING HISTORY
Berkley premium edition / July 2013

PRINTED IN THE UNITED STATES OF AMERICA

10 9 8 7 6 5 4 3 2 1

Cover design by Pyrographx.

ALWAYS LEARNING   **PEARSON**

This book is dedicated to the following UFO authors and researchers who provided me with their time and their knowledge during the research and writing process:

Nick Redfern, Jerome Clark, Christopher O'Brien, Norio Hayakawa, Bill Birnes, Nick Pope, Andrew Hennessey, Nikolay Subbotin and Michael Kinsella

ACKNOWLEDGMENTS

Thanks to Tom Colgan and Amanda Ng at The Berkley Publishing Group; Dominick Abel, Jacob Boucher, Larry Stone, Bob Messia and BP for the laughs; Charles Ketchum for his spiritual guidance and Chase for letting me on her radio show. Special thanks to my Spook friend and his Spook friends, without whose help this would have been an entirely different book, and thanks to Sky Club for doing the sound track.

Special thanks to the Mutual UFO Network (www .mufon.com) and the National UFO Reporting Center (www.nuforc.org).

A very special thanks to my best friend ever. She knows who she is.

FOREWORD

by NICK REDFERN

One of the biggest puzzles concerning secret military-, government- and intelligence-based facilities is this: Why do so many of these classified installations seemingly attract far more than their fair share of UFOs?

There are, I believe, two prime possibilities.

The first is that if the stories of crashed UFOs in the hands of officialdom are valid, then it may be the case that what we are seeing in the skies above these bases are (a) attempts by military project personnel to test-fly captured UFOs, (b) attempts to test-fly their own versions of alien technology or (c) a combination of both.

There is another more ominous possibility, however.

Again, if UFOs really have crashed to earth, perhaps

the nonhuman intelligences behind the phenomena are aware that their craft and dead (and possibly even living) comrades are held at these fortified bases, and they are testing the defenses of such places for the day when they plan to claim back what is theirs.

If that is the case, then how might we best go about trying to vindicate such theories and scenarios?

Well, a very profitable approach would be to do exactly what Mack Maloney has done—namely, to use investigative journalistic skills to chase down just about every piece of available data and evidence on unusual and classified installations around the world and then present it to one and all for consideration.

Indeed, in the pages that follow, as you will quickly come to learn, Mack has expertly negotiated the globe in search of the truth behind such off-limits facilities and the amazing secrets they so jealously guard. . . .

—Nick Redfern

PART ONE

Secret Places in America

1

Beers with a Spook

It took three weeks to arrange the meeting.

My friend who works with the two most prominent U.S. intelligence agencies agreed to have a few beers and talk about a subject that makes him very uncomfortable: UFOs.

Our conversation took place in a waterfront bar north of Boston. It was scheduled for a weekday afternoon when we hoped customers would be few, and the place was nearly empty when we arrived.

The weather was oddly appropriate for such a meeting. Sunny and warm over most of the area, it was damp and gloomy where we were, the fog hugging the water like a stratus cloud. Not really typical for a mid-spring day.

I'd previously given copies of *UFOs in Wartime* to my friend and several of his colleagues. Their response had been positive but muted—not that I was expecting anything more from people in their business. At an earlier meeting I'd asked my Spook friend if he'd ever seen any evidence that the U.S. government knows what UFOs are. His answer was no. But when I asked if he'd be able to *tell* me if he'd seen such evidence, his answer was a slightly less emphatic: "No."

That's how it is in the U.S. intelligence community when it comes to UFOs. No, we've never seen anything like that—but no, we couldn't tell you if we had.

I explained the idea behind my new project: a book examining unusual places around the world, many of them secret military bases like Nevada's Area 51, that seem to have an inextricable connection to UFOs. Some of these connections are truly puzzling, and some so outlandish they border on ridiculous. But the most interesting ones fall somewhere in between. As my Spook friend was well aware, just about every secret base that we know about has some kind of link to UFOs. The book would try to find out why.

With this in mind, I had two simple requests for him.

Could I keep him updated on my progress, just as a way of not veering too far off course? He agreed I could. Second, once the first draft was complete, would he read it over and tell me his thoughts at a second meeting? Again, he agreed.

It was raining by the time we left the bar, adding to the murk along this little piece of coastline. Everything seemed black-and-white—except off to the west, where there was an opening in the overcast that was shaped like a keyhole. Beyond that opening was a patch of bright blue sky.

The writing of this book began the next day.

2

Early Secrets

It's no surprise that finding history's first "secret base" is not an easy task.

By definition, secret bases *should* be hard to find, whether on Google Earth or in the pages of a history book.

When two parties go to war, it usually benefits each side to keep the size of its true strength from the other. This is also true when preparing for hostilities, or in times of peace between wars. This is where secret bases come in, places where you can keep your latest weapons, your newest equipment, even your best soldiers, out of view.

Finding the first secret base in history might actually be impossible simply because it may have come and gone without many noticing.

But a good candidate we *do* know about is a place called Megiddo.

For more than four thousand years, and maybe as far back as seven thousand years, this city, located in what is now northern Israel, saw empires, kings and pharaohs come and go like so much dust in the wind. So many military installations had been built there, one on top of the other, that what began as a village at ground level eventually became a small mountain fortress. This upward expansion was so unusual that around 3000 BC, when cities first began to develop around the ancient world, Megiddo was already a metropolis.

Because Megiddo commanded a key part of the strategic road that linked Egypt to Assyria, countless battles were fought for control of it. Military historians agree that more important battles have been fought in or around Megiddo than probably any other place on Earth.

In the words of Pharaoh Thutmose III, "The capture of Megiddo is the capture of a thousand cities."

But how can such a historically prominent place be considered a secret base?

For much of its history, Megiddo was surrounded by

stone fortifications and earthen ramparts. At its height there were also many carefully divided quarters within its walls, places set aside for Megiddo's royalty and citizens but also for its military.

Because of these circumstances, a case can be made that at Megiddo, for the first time, there was a conscious effort to hide the true strength of its resources. For instance, for a long time the water supply for the city lay in a spring inside a cave located outside the city's walls. But at some point, no one is really sure when, it became clear that in times of war or a siege, the city's water supply had to be secure. So a subterranean water system was dug down through rock to this spring, allowing Megiddo's residents to collect water without being exposed to the enemy.

But then the city went a step further. They built a thick wall at the entrance to the cave where the spring was located, blocking it from the outside. This wall was then camouflaged by dirt and flora in order to hide the source of water from Megiddo's enemies.

So not only was the city's key resource secured, but it was also concealed.

In order to utilize the water from this spring, the people of Megiddo built massive stone troughs and then, in the middle of the fortress, a giant watering pool was constructed. This in turn allowed large numbers of horses to be kept inside the walls, out of sight. It's estimated that Megiddo's stables may have accommodated up to five hundred horses at one point, a huge number in ancient days. More telling, other structures within the fortress were said to have housed hundreds of the newest and best in battle chariots. Again, out of sight of any potential enemies.

This fortress city has another historic aspect to it. Because of its vertical evolution over the years, Megiddo eventually found itself atop what in Hebrew is called a *har*, or hill. Hence, the fortress's real name was Har-Megiddo, or . . . Armageddon, the place the Bible says will be the scene of Man's final battle against Evil.

For our purposes, though, we can call Megiddo a "secret base," because it intentionally kept its secrets hidden, away from the enemy's prying eyes.

But at the same time, we must recognize it was a secret base that was very much hiding in plain sight.

10

3

The Mysterious Mountain

Love and UFOs

In the middle of the Nevada desert, about seventy miles northwest of Las Vegas, there's a patch of salt flats called Papoose Lake.

There's a mountain close by, impossible to see from public land but indistinct and ordinary looking by all reports, and no different from the dozens of other mountains in this part of Nevada.

Yet some claim that contained within this mountain is nothing less than the epicenter of the "New World Order"—not quite the one proclaimed by President George H. W. Bush back in 1990 but a deeply secret, internationally controlled shadow government of the same name whose goal is nothing less than domination of the entire planet.

But then, there are others who believe this mountain is actually a garage for crashed UFOs—flying saucers the U.S. military has recovered and kept hidden from us over the years.

Still others are convinced it's a kind of extraterrestrial holding cell, a place where space aliens are kept under house arrest. Or that it's the entrance to a tunnel that stretches all the way to Las Vegas. Or that it's the hideout of a highly illegal CIA assassination team that's run by ex–Vice President Dick Cheney.

Then again maybe it's just a mountain—and the only reason these tales even got started has more to do with a love story gone wrong than UFOs or aliens or a secret government taking over the world.

Only one thing is certain: This mountain, known as S4 (as in "Sector 4"), is located in an interesting neighborhood. Because just fourteen miles away is the Air Force Flight Test Center at Groom Lake, the secret U.S. air base better known as Area 51.

Not many people know for sure what's going on there either.

The Beginning of Bob Lazar

Although it might not seem like it now, Area 51 was relatively unknown to the general public until the late 1980s.

Of course, those high up in the U.S. military and the country's intelligence services were familiar with the place. They called it "Dreamland," "Paradise Ranch" or simply Groom Lake. It was here that the United States quietly tested top-secret aircraft like the U-2, the SR-71 and various stealth airplanes. But for most of us, Area 51 just didn't exist, not yet.

All that changed one night in May 1989, when a Las Vegas TV station ran a story about a young, unnamed physicist who claimed he'd worked at both Area 51 and its extension at S4, and that he'd seen the remains of damaged UFOs there, machines not of this earth, kept under lock and key by the U.S. government.

A few weeks later this physicist appeared on TV again. This time he was identified as Bob Lazar, age thirty, of Las Vegas, and he repeated his assertion that he'd worked at Area 51 and that the U.S. government had at least nine UFOs hidden away inside the S4 complex nearby.

Suddenly Area 51 became the most famous secret base in the world.

* * *

Lazar's story went like this: As an employee of Area 51, he was regularly taken by bus to S4, located about twenty minutes away from Area 51 / Groom Lake.

According to Lazar, S4 was carved into the base of Papoose Mountain. He described the entrance to the facility as having nine sloped doors with a sand-like camouflage texture on them, allowing them to blend in with the rest of the mountain.

Within the mountain itself was a huge man-made cavern containing nine hangars—thus the nine doors. A damaged UFO could be found inside each hangar.

Lazar says he was put to work on these alien craft, studying their propulsion systems and trying to figure out exactly how they worked and how the technology could benefit the U.S. military. In other words, he was asked to help reverse-engineer the UFOs.

In later interviews, Lazar went into detail on what these UFOs looked like. One saucer-shaped craft in particular he called the "sport model." He claimed its exterior skin was metallic and the color of unpolished stainless steel, and that it stood upright on tripod legs. Its entry hatch was located on its upper half with the bottom portion of the door wrapping around the center lip of the disc. Basically a UFO as seen through the eyes of a Hollywood prop department.

Lazar said the saucer's interior was divided into three levels. The lower level contained three "gravity amplifiers" and their "wave guides." The disc's reactor was located in the middle level, directly above the gravity amplifiers. This level also housed the craft's tiny control consoles and seats, items just too small and too low to the floor to have been built for adult human beings.

Lazar claimed the craft's reactor used fuel that did not occur naturally on our planet. Its main ingredient was a superheavy element with an atomic number of 115, something that does not appear on any Earthly periodic table.

Finally, Lazar said that even though he worked on the discs for several months, he never learned how or where any of them had been recovered.

Nor did he ever encounter any extraterrestrial beings while he was at S4.

The Tale of J-Rod 52

This was not the case for a man named Dan Burisch.

According to his own story, which came out a few

years after Lazar's revelations, Burisch was a microbiologist who worked for both Naval Intelligence and the Defense Intelligence Agency from 1991 to 1996. He claimed he served at Nellis Air Force Base—the massive military installation located right next to the famous Las Vegas Strip—and that at some point he went to work at Area 51 / Groom Lake and then at S4.

While working at S4, Burisch said he was able to do something Lazar had not: meet an actual alien. In fact, Burisch was asked to take tissue samples from an ET being kept at S4 named "J-Rod 52." Described as a typical "Gray" extraterrestrial (those familiar smallish aliens with large, bald heads, sickly skin and giant eyes), J-Rod 52, Burisch said, was one of two extraterrestrials that survived a UFO crash in Kingman, Arizona, in the early 1950s.

Burisch said that J-Rod 52 was under the weather when he first met him. Over the next two years, though, the two became close as Burisch continued, it is assumed, taking tissue samples from his alien friend.

During this time, J-Rod 52 revealed to Burisch that his alien race had actually inhabited Earth many thousands of years ago, before being forced to leave. The reasons for their mass departure were many. They included a great global catastrophe caused by massive solar flares, a violent shift in Earth's poles, and extensive crumbling of the planet's crust. This evacuation forced J-Rod's kind to wander among the stars. But now they were back to retrieve what Burisch said J-Rod called a "lost genetic factor" from the human race, and presumably make friends when possible.

Burisch's intergalactic bromance came to an end, though, when an American "satellite government" decided J-Rod 52 might be more useful if he helped them better communicate with his home planet.

To this end, J-Rod 52 was taken to a "natural star gate" in Abydos, Egypt. Once there, Burisch himself pushed his alien friend through the star gate, where he disappeared, never to be seen again.

Illuminating the Illuminati

How S4 became connected to the New World Order requires an even further leap of faith.

Rumors of the "Illuminati" secretly controlling world events go back centuries. No surprise then that some conspiracy enthusiasts (author and filmmaker Anthony Hilder, for one) think these enlightened boogeymen—who are just pouring out of places like Harvard, Oxford and Yale these days—have a hand in what goes on at S4.

One scenario posits that the strange discs located at S4 might actually be of earthly manufacture, and that there are a lot more of them than just the nine Bob Lazar saw. The UFOs' raison d'être is they will take part in a massive Illuminati-orchestrated plot to stage a fake but realistic reenactment of Orson Welles's infamous 1938 radio broadcast of *The War of the Worlds.* Why? Because, so the theory goes, when people around the world see hundreds of the S4-manufactured UFOs zooming overhead, they will be convinced Earth is being attacked by extraterrestrials and panic will set in, just as it did during the Welles radio broadcast. This will lead the masses to urge their individual governments to give up sovereignty and join the rest of the people of Earth in fighting off the outer space menace.

Of course, when every country's sovereignty is gone, the ruse will be revealed. But it will be too late.

The Illuminati will have their dream of a "One World Government," and we will enter the era of the New World Order.

"It Should Be Wet"

With all this drama surrounding S4, one has to wonder what kind of people are guarding the place.

The Area 51 / Groom Lake facility is itself guarded not by the military but by employees of the Wackenhut Corporation, also known as G4S Secure Solutions. These no-nonsense heavily armed guards continuously patrol the borders of Dreamland, operating behind signs that threaten the use of deadly force should any unauthorized persons try to enter the base.

But apparently when it comes to S4 itself, some claim nothing less than highly trained, cold-blooded SEAL assassins in the employ of the CIA serve as sentries.

According to controversial scholar and *Exopolitics* author Michael Salla in two articles he wrote for Examiner.com, whoever is running S4—the U.S. government, a shadow government or the Illuminati—is brutally selective about who protects its secrets as well as its elevators.

According to Salla's articles, one such sentry was a man named O'Ryan or O'Hennessy (take your pick), who joined the Navy (or the Marines—again, the story varies), then the SEALs, and was *then* recruited by the CIA to become a so-called Black SEAL. This landed

"O'Hennessy" in a top-secret and highly illegal assassination squad overseen by ex–Vice President Dick Cheney. This assassin squad targeted not only the usual enemies of the United States but also allegedly "disloyal" U.S. citizens as well.

Back in the 1960s, CIA assassination operatives supposedly used the phrase "it should be wet" as a code for troublesome people who had to be terminated. Apparently that was O'Hennessy's job. He claimed to have whacked eighteen people for God and country, at least one of them an American citizen, for reasons unknown. But because it's illegal for the CIA to assassinate anybody these days (drone strikes notwithstanding), once the death squad's button men got too hot, so the story goes, they were reassigned to guard the elevators at S4.

According to Salla, O'Hennessy worked on the second level of the S4 complex, which at the time was reportedly made up of four descending levels in total. While O'Hennessy was one of about six dozen people assigned to level two, his previous assassination work didn't carry enough juice to get him clearance to visit the lower levels of three and four, reportedly where all the *really* secret items were located.

O'Hennessy did get a chance to look around the upper part of S4, though, and in some respects his story differs from that of Bob Lazar. The former Black SEAL described level two as a hangar-type space with seven retrieved flying saucers resting in individual bays, with three bays being empty.

The guard claimed one or more of the flying saucers was flight-worthy and would occasionally be raised to the surface for a flight test, or maybe just a joyride. But the people running S4 were careful not to allow this to happen when Russian or other countries' spy satellites were passing overhead.

Again, according to Salla, O'Hennessy got to see his alleged boss, Dick Cheney, on at least one occasion, as he claims the then secretary of defense visited S4 in 1991 and that O'Hennessy was on duty that day.

People who were later in touch with O'Hennessy claim he showed them photos of what appeared to be extraterrestrial vehicles and extraterrestrial bodies in glass tubes plus a picture of Dick Cheney standing on a balcony looking down on them all.

But why would a super-SEAL like O'Ryan/ O'Hennessy, someone who would have required a mountain of high-level security clearances, come forward to spill his story to the world? Supposedly it happened after he discovered he'd contracted a serious disease while on duty, something he felt was intentionally given to him by the people running S4.

Salla says that after providing a number of interviews, some of which can be found on YouTube, the mysterious Black SEAL simply disappeared.

The UFO VP?

How credible is O'Ryan/O'Hennessy's rather incredible story?

As it turns out, it's not totally unsubstantiated. As reported in the *New York Times* in July 2009, there *was* a secret CIA death squad in existence in the years after 9/11, its aim being to illegally assassinate people overseas. Then Vice President Dick Cheney, if not actually running it, was at least aware of the hit team and sought to keep knowledge of it from Congress.

Cheney also has an interesting history with UFOs. On April 11, 2001, he was asked on the *Diane Rehm* PBS radio program if he'd ever been briefed about UFOs. Cheney's Spook-talk reply surprised a lot of people.

He said, "If I had been briefed on it, I'm sure it was probably classified and I couldn't talk about it."

Going Full Circle

Which brings us back to Bob Lazar.

Who was he really?

Bespectacled, with a bookish air about him, he certainly looked the part of the physicist done wrong by his country, say most people who met him.

But was he even a physicist at all?

According to Lazar, he attended both MIT and Caltech, receiving a master's degree from MIT. He claimed to have studied physics at Caltech and said he later worked as a physicist at the Los Alamos National Laboratory.

However, investigators checking with both MIT and Caltech could find no record of Lazar's attendance. But Lazar's supporters argue that this was simply the result of the government's plot to alter his records and discredit him.

Lazar claimed he abruptly left his job at S4 after he discovered his birth certificate had disappeared as well. He said government operatives had expunged his hospital birth records along with his college transcripts and other employment records.

This began his fear, he said, that the U.S. government was going to, in effect, make him disappear—

sanitize his identity and perhaps even kill him because he'd served their purpose. Lazar further claimed he'd had his life threatened and was shot at.

It was for this reason that he went public with the television broadcasts and, later on, with many interviews and personal appearances as well.

A Weird House in Vegas

Norio Hayakawa is a noted UFO researcher and author. Not only has he written extensively about Area 51 and S4 but he also had personal dealings with Bob Lazar.

"He was an enigmatic person," Hayakawa told this writer in an interview. "And very mysterious."

To illustrate his statement, Hayakawa relayed the following anecdote.

Hayakawa went to a place where Lazar said he lived, a house in the Las Vegas area. The occasion was for yet another interview, but no matter where Lazar went in the house, he always had two strange men with him, even sitting on either side of the couch with him while he and Hayakawa talked.

"When Lazar would get up to go to the bathroom, these two guys would go with him," Hayakawa said. "That was strange."

The house was strange, too, Hayakawa said. It didn't look like it was being lived in on a daily basis.

At one point, Hayakawa managed to get a look at Lazar's W-2 tax form. Lazar had previously claimed he'd worked at S4 for several months. Yet according to Hayakawa, his W-2 indicated only about $985 in return for five days' work for the government. That comes out to be about twenty-five dollars an hour—not bad for the late eighties but certainly not the paycheck of a top-level, high-security-clearance physicist.

Hayakawa also investigated the Social Security number on Lazar's tax form and discovered it was not his but actually belonged to Lazar's former wife.

Which leads to what might be the most intriguing question of all: Could a simple romance gone wrong have been behind Lazar's S4 saga?

The Real Story?

Bill Birnes is a well-known UFO researcher and star of the TV show *UFO Hunters*.

Regarding Bob Lazar, Birnes told this writer bluntly: "The real Lazar story is so weird you couldn't make it up."

Birnes said that Lazar did indeed work at Area 51,

but his wife became suspicious that he was away from home so much and couldn't tell her what he was doing during that time. She suspected he was cheating on her and that the story of doing secret work for the government was just a cover. So she confronted him with it.

Hoping to save his marriage, Lazar felt he had to prove to his bride that he *was* working at Area 51. So he took her out to a certain place in the Nevada desert on a certain night, and once there, just when he said it would happen, she saw something very strange in the sky, something she took to be a UFO. Understandably, she was astonished. This seemed to be proof positive of what Lazar had been telling her. If not, how would he have known when one of Area 51's "UFOs" was going to take flight?

But then Lazar started bringing other witnesses to this same spot and showing them whatever it was flying around out there, and eventually he got caught. The security people at Area 51 pulled his clearance and told him he was in deep trouble. They fired him and indeed sanitized his records.

In retaliation, Lazar ran to the media and told the S4 story, and it blew up from there.

Disinformation, Please . . .

According to Hayakawa, Lazar doesn't do interviews anymore.

But that doesn't mean life has calmed down for him since leaving the public eye. Hayakawa revealed, for instance, that in 2006 Lazar's house was raided by DEA and FBI agents who accused him of selling material that could be made into weapons. The case was later dropped.

Hayakawa says Lazar eventually moved to Michigan. "He was selling scientific equipment," Hayakawa said. "Items that you can't get easily. His specialty was dealing with defense contractors."

Dealing with defense contractors? Doesn't that sound like someone who's "connected" in some way?

Which leads to another question that's been asked before: Was Lazar a disinformation agent all along, someone working *for* the government?

After all, it's been reported that before he first appeared on TV, Russian satellites would routinely photograph Area 51 about three times a month. But after Lazar talked about the crashed discs and so on, the Russians began photographing the Papoose Lake region and S4 almost daily.

So was Lazar's ultimate role to misdirect the Russians into wasting their time and resources photographing nothing more than an ordinary mountain in the middle of Nevada?

Could it be as simple as that?

"He had a significant personality," Hayakawa said. "But whether that personality was created by someone else? I don't know."

But whatever Lazar was up to, he never wavered from it.

"He never changed his story from day one," Hayakawa concluded. "That doesn't mean that what he said was true. But he stuck with it."

Shooting Holes in S4

Government plant or not, there are fundamental problems with Lazar's story, aside from missing birth certificates and nonexistent college degrees.

For instance, Lazar said he was bused from Area 51 to S4. Yet, as many investigators have pointed out, there are no visible roads that go from Groom Lake to Papoose Lake.

Then, there's the basic question about the rumored

S4 itself. How does one build a place like that? Coring out a mountain, turning it into a facility that would have to be ultra-high-tech, if indeed it was a repository for something as earth-shattering as crashed UFOs? And do it all in total secret?

To construct such a facility would take thousands of workers and tons of earthmoving equipment. The project would require massive amounts of power, generators and hydraulic tools and fuel to run them. And wouldn't you do the work at night, so those pesky Russian satellites wouldn't see you from one hundred miles up? If so, think of the massive amount of lighting that would be needed to create something on the scale of S4.

Then there's the problem of the dirt. Where does one put the hundreds of thousands of tons of dirt and debris that would result from digging into the side of a mountain to create such a huge cavern within?

This debris could be loaded onto trucks and hauled away, but where would they haul it to? And on what roads, if there are none stretching between Area 51 / Groom Lake and the supposed site of S4?

The whole idea of a vast underground base, costing billions, built in secret, by tens of thousands of workers, with no visible signs or evidence of its construction, all to hold crashed UFOs or whatever other exotic government secrets there are . . . it seems a bit far-fetched.

Until you learn about a place called Mount Yamantau.

"We Don't Have a Clue. . . ."

On April 16, 1996, the *New York Times* reported that a mysterious underground complex was being built by the Russians in the Ural Mountains. The project was so big, tens of thousands of workers were involved and an entire highway and new railroad extension had to be built in order to service it.

U.S. intelligence sources believe the Russian government has pumped more than $6 billion into the Mount Yamantau complex, a cavern that spans some four hundred square *miles*. One theory suggests that in the event of a nuclear war, the Russian leadership would head for Yamantau, a place where they could survive until the aftereffects of such an apocalyptic conflict settled down.

But there might be more to it than that—especially considering, as we will see, other unusual UFO-related events going on inside Russia.

Could Yamantau be another Area 51 located deep inside a mountain? A "Russian S4"?

No one really knows.

But back in 1998, in a rare public comment, then commander of the U.S. Strategic Command General Eugene Habiger said these chilling words about Mount Yamantau:

"It's a very large complex that we estimate has millions of square feet available for underground facilities. [But] we don't have a clue as to what the Russians are doing in there."

A hollowed-out mountain in the middle of nowhere? Built by tens of thousands of workers in relative secrecy? Millions of square feet for underground facilities, their true purpose secret and unknown?

It begs the question: If the Russians did it, why wouldn't we?

4

The San Luis Valley

Tales from Mysterious Highways

On October 16, 1996, around ten p.m., a man driving his pickup through Colorado's San Luis Valley stopped at the side of the road to relieve himself.

Two female relatives were traveling with him; they were on their way to a wedding the next day.

But when the man got out of his truck, something strange happened. He noticed his shadow on the ground. This wasn't right; the sun had long since set and it was a moonless night.

The man looked up and to his astonishment saw a

gigantic UFO hovering not fifty feet above his head. The craft was saucer-shaped, had dozens of blinking lights circling its rim and was making no noise. Yet all around him the air itself seemed to be rumbling.

The man yelled a warning to his passengers; they saw the craft and began screaming hysterically. Though he rushed back to his truck, the man just couldn't take his eyes off the thing.

Suddenly the UFO shot straight up, to maybe a mile or so. But then just as quickly it came back down and resumed hovering over the pickup truck.

Finally it moved away, heading north on a course parallel to the highway. Over the strenuous objections of his passengers, the man started his truck and began pursuing the UFO.

After driving about five miles, he caught up to the strange craft. But suddenly the UFO was back over the pickup truck. Though his female passengers begged him not to, the man climbed out of the truck again and studied the UFO for a second time.

At this point the UFO turned off all its lights and moved away again. The pickup's passengers heard another low vibration, which they were certain was emanating from the craft.

Then, the darkened UFO drifted out over the desert and finally disappeared.

* * *

On September 3, 2000, three men driving through the San Luis Valley were startled to discover a strange red light following their car.

It wasn't another car, motorcycle or truck. It was just a single disembodied red light—and it was right on their tail.

The driver hit the gas. The mysterious light sped up as well. A chase ensued, the driver trying to shake the strange light, only to have it keep pace. Even driving like a madman through a series of S-curves in the roadway didn't work. The UFO stayed right with him . . . until it abruptly disappeared.

On October 24, 1998, a motorist driving through the San Luis Valley spotted two large glowing white objects in the sky moving at a high rate of speed. At first, the motorist thought they were meteors, but this was unlikely as the sun was still out. Three days later, another motorist driving along the same highway saw a bright light disappear into a cloud—and not come back out.

Five months earlier, witnesses on the same highway saw what they described as "Hollywood-type spotlights" pointing up into the sky from the tops of the mountains in the distance. These bright lights, described as both white and red, appeared to be blinking in sequence. A few days later, other motorists saw the same strange sight—large spotlights along the mountaintops, again blinking on and off in sequence.

How Weird Is My Valley?

Colorado's San Luis Valley has been called the "New Area 51," but that's really not an accurate designation. Something not only very strange but also very unique is happening in this place, stranger even than the most fantastic stories to come out of Groom Lake and its environs.

Located south of Denver, the "SLV" is 75 miles wide and 130 miles long. Encompassing about 8,000 square miles, it sits almost a mile and a half above sea level and makes up a large part of southwestern Colorado. With the Sangre de Cristo Mountains on the east and the San Juan Mountains on the west, this alpine valley is a beautiful part of the American West, postcard-perfect in many places.

Yet UFOs of all shapes and sizes have been seen flying

around the SLV for many years. These include saucers both big and small as well as cigar-shapes; needle-shapes; rocket-shapes; pencil-shapes; sausage-shapes; fireballs; huge triangles; low-flying red lights; alternating green, red and white lights flying in spherical formations; light gray objects; metallic spheres; globes of light of all different colors; UFNs (unidentified flying noises); wingless airplanes; elongated cylinders and at least one moonsized object with a visible tail.

And that's just a *partial* list.

No surprise, the SLV has the highest incidence rate per capita of UFO sightings anywhere in the United States, according to the Computer UFO Network database.

What's going on here? Christopher O'Brien is a paranormal investigator and author and an expert on the SLV. As of this writing, he's authored five books, including the essential *Secrets of the Mysterious Valley*.

O'Brien lived in the sparsely populated valley for years and still maintains a network of sky watchers and associates there. In that time he compiled a database of strange activity inside the SLV that runs for hundreds of pages. Probably no one knows more about the unexplained in the SLV than O'Brien.

Yet even he's baffled.

"You ask, 'What's going on in the valley?'" he said during an interview with this writer. "I don't have an easy answer for that."

What Could Be Stranger Than a UFO?

When it comes to oddities in the SLV, it just begins with UFOs.

Consider the "prairie dragons."

Undulating forms that seem to swim in the air a foot or so off the ground, prairie dragons look like giant sea slugs, but most people only see them out of the corner of their eye. They don't leave tracks and they don't make any noise, but most interesting, animals always seem to recognize their presence.

Then, there are the "shadow people." People in their homes will see shadows of figures on their walls, but no one, or no thing, is in the room causing them.

The region also has ghosts, especially at a place called the Silver Cliff Cemetery, located just east of the SLV in Silver Cliff, Colorado, population four thousand. Silver Cliff residents see strange lights over this graveyard all the time. In fact, the entire town shut off its lights one night just to see if the unexplained graveyard illumination was coming from some kind of odd reflection of someone's front-porch light. The town went completely dark, yet the graveyard lights still appeared.

The SLV experiences some very strange weather, too. Not only can storms form and strike quickly but sometimes the fog is so thick and so low to the ground it resembles cumulous clouds.

"But it's even stranger than that," O'Brien reports. "At times the fog seems to be 'encased.' It will cover a pasture but stop right at the fence lines. Other times it will make weird shapes. I personally saw an enormous saucer-shaped cloud floating above the Great Sand Dunes. It was so perfect, I took a picture of it. But not a couple minutes later, the thing had completely vanished."

These days, cattle mutilations practically make up a branch of UFO study on their own. But in the SLV, it's not just cattle that are being victimized. O'Brien reports that horses, sheep, deer, elk, pigs and even coyotes have been found mysteriously killed and cut up, with certain parts missing, other parts still intact. Mutilated but not butchered.

"There was even a case of a javelina mutilation," O'Brien said, referring to the hairy piglike animal. "It was found up in the branches of a tree, all cut up. The problem is, there aren't supposed to be any javelinas in the valley. They're found in Arizona and other places, but not in the SLV. There's no reason one should be up there."

Residents of the SLV also hear and experience something like the "Taos Hum," a nauseatingly low drone or vibration.

"It resonates through your whole body," O'Brien said. "You don't really hear it; it's more like you feel it."

In addition to the more typical UFO shapes, people in the SLV have also seen what are best described as wingless airplanes. O'Brien has seen one himself. "They are huge and look more like the space shuttle," he said. "But they look like a space shuttle with no wings."

And lately, there have been reports of flying humanoids in the valley. A cross between Dracula and Mothman, these strange aerial creatures have been seen by many reasonable and reliable people, O'Brien says, including policemen and municipal workers.

And if that wasn't enough, there have also been many sightings of Bigfoot in the mountains and foothills surrounding the SLV.

BEYOND AREA 51

How "Hot" Is Hot?

So, is the SLV the hottest of hot spots when it comes to UFOs and assorted paranormal happenings?

Actually, O'Brien uses certain criteria to define a "hot spot." The area in question has to fulfill each of these five points to qualify:

1. Unusual geophysical properties—At points on the globe where huge tectonic plates come together, the Earth's magnetic field might have a very strong force on one side and a weak one on the other, leaving a seam or a rift. A lot of UFO sightings tend to be made along these rifts. The SLV is close to the Rio Grande Rift Valley, one of the largest rifts in the world.

2. A culture of Native American sacredness—The SLV is the only area in North America that has been shared by three different regional groups of Indians. More than a dozen distinct tribes have come to the valley over the centuries for things like vision quests. Most important, though, is that Blanca Peak, the Sacred Mountain of the East to the Navajo and Apache, is located here. This is a place where, some tribes say, "all thought originates." And near the

41

SLV's Great Sand Dunes is the location of the *Sipapulima* (or "place of emergence"), according to Tewa and other upper Rio Grande Indian lore.

3. Multiple waves of unusual events—Not just flying objects, but also aberrant behavior, animal mutilation, strange weather and so on are frequently reported in hot spots. Strangeness in the SLV comes and goes; the 1990s in particular were an intense period of activity and thus provide some of the best stories. Also interesting is that O'Brien reports that when a wave is just starting up in the SLV, it seems to serve as a precursor to strange events elsewhere— like hundreds of miles away in the Black Hills of the Dakotas. Is this the paranormal equivalent of what Einstein called "spooky attraction at a distance"?

4. Close government proximity—The SLV is only about eighty miles from Colorado Springs, a place crammed with classified military installations. Plus, the La Veta military operations area, a gigantic swath of land just outside the valley to the east, run by the U.S. Air Force, is set aside for low-level military flight training.

5. Attempted military expansion into the area—The fifth element of a hot spot is true in the case of the SLV—the U.S. Air Force is constantly trying to buy up or take over more land inside or near the valley.

Secret Bases and Black Ops Nearby

"A lot of what people see in the SLV are probably black operations being done by the military," O'Brien said.

Again, Colorado is home to a number of classified military bases. Many of them are close to Colorado Springs, which is only a few minutes' flying time from the SLV.

At least three of these nearby bases are involved in highly secret operations. Schriever Air Force Base operates more than 170 of America's spy satellites and is also the main control point for the Global Positioning System, what we know as GPS. Peterson Air Force Base is the home of the North American Aerospace Defense Command (popularly known as NORAD) and the Air Force Space Command, whose 21st Space Wing provides missile warning and space control to U.S. combat forces worldwide. Buckley Air Force Base is involved in controlling America's spy satellites as well, but the base also hosts the Colorado Air National Guard's 120th Fighter Squadron and its F-16C fighters, among other air units. The Colorado National Guard also operates CH-47 Chinook, UH-1 Huey and UH-60 Blackhawk helicopters out of Buckley.

Just outside Colorado Springs is Cheyenne Mountain, the home of the Cheyenne Mountain Air Force Station, NORAD's former location. Known by its code name "Crystal Palace" during the Cold War, the Cheyenne station now serves as a backup to NORAD.

In case NORAD has to go back underground, it would return to Cheyenne Mountain.

It's widely suspected that the U.S. military tests secret aircraft inside the SLV, just as it does at Groom Lake and other places two states over in Nevada. And maybe that explains things like wingless airplanes.

But what about the instances where people in the valley have seen military jets buzzing around UFOs?

"I've had law enforcement witnesses report they've seen a huge black triangle-shaped object being 'escorted' by F-15 fighters," O'Brien said. "And I've had a report of witnesses seeing a small black triangle being *chased* by F-16s."

O'Brien himself has seen such things. One night he got a call that two UFOs were heading in his direction. He went outside his house just in time to see a pair of large orange balls moving at high speed, heading south down the center of the SLV at treetop level. Exactly eight minutes later, two F-16s roared over as well.

"The jets were obviously pursuing these objects," he

said. "The long trail of flames coming from their afterburners told me both planes were at full military power."

Strangely Routine

Strange events occur in the SLV so often, the residents almost think of strangeness as the routine.

In fact, unusual things have been happening in the SLV for a long time. O'Brien cites an entry found in the diary of the first Spanish governor of the New Mexico Territory in 1777. The governor mentions that his soldiers have witnessed lights and heard an unidentified humming in and around the SLV's Blanca Peak.

"So many unusual things happen there," O'Brien said, "that people don't think it's any big deal. No matter what it is, they're used to it. That's why I suspect that a lot of events are never reported."

Still, some stand out.

What follows are some of O'Brien's most baffling SLV incidents.

The Case of Snippy the Horse

On the morning of September 8, 1967, a rancher in the San Luis Valley went out to water his horses. He owned three but on this day only two were waiting for him. There was no sign of the third, a female Appaloosa named Snippy.

As the horses had a lot of open space on which to roam, the rancher waited a full day for the third horse to show up. But once twenty-four hours had passed, he set out to look for her.

After about an hour of searching he spotted the horse in a meadow not far from his main house. It was a gruesome discovery. From the tip of its nose down to its shoulders, the horse was nothing but bones. All of the skin, muscle and tissue from the neck and skull were missing and the exposed bones were bleached white, as if they'd been out in the sun for years.

Though horrified, the rancher made note of the immediate surroundings. He determined from tracks he'd found that his three horses had been running at full speed whenever the incident had happened. He surmised that the dead horse must have been cut off from the other two that eventually returned to his ranch house.

But the strangest thing of all was his discovery that the dead horse's tracks continued for several hundred yards before they inexplicably stopped—while still in full gallop.

The rancher later discovered a number of burn marks on the ground near the horse's remains. Also found were what several people described as gigantic horse tracks measuring eighteen inches wide and eight inches deep.

Though various investigators and media members looked into the strange incident, just what happened was never determined. One thing *was* certain: The horse wasn't killed and then partially eaten by scavengers. The condition of the carcass showed no evidence of this.

The case still baffles people of the SLV today.

The Mysterious Yellow Helicopter

On June 5, 1980, a seven-hundred-pound prize bull had been put in a small pasture by its owner. This pasture was about a quarter mile from the owner's ranch house.

Just around dusk on that same day, the owner's family was sitting down to dinner when they heard the sound of a helicopter approaching. The sound grew until the copter went over their house, flying very low. While the family was used to seeing utility-company helicopters flying around, the nearest set of power lines was located more than three miles away. That's why the family thought it unusual that a helicopter should be flying so low right over their domicile.

A few minutes passed, and then the family heard the helicopter again. But this time it sounded like it was taking off from the vicinity of the bull's pasture. The family went outside and saw the aircraft for the first time. They described it as an old-fashioned, two-man, whirlybird-type helicopter, mustard-yellow in color. And indeed it was taking off from the pasture where their bull was located.

The helicopter flew right over the ranch house again, no more than forty feet above their heads, before disappearing to the north.

The next morning the family went down to the pasture and found their bull was dead. Its sex organs and eyes were gone, its rectum had been cored out and a plug was missing from its flanks. Inexplicably, a horde of flies that had been eating one part of the carcass were also dead and scattered around the body.

Angry that their bull had been killed, the family called every aviation facility in southern Colorado, hoping to locate the home base of the old-fashioned heli-copter. Trouble was, no one had seen anything like the whirlybird-type aircraft the family described. What's more, the family was told that particular model, which was from the Korean War era, not only was extremely rare but would have been extremely expensive to fly be-cause of its age. Plus, due to its high fuel consumption and decades-old avionics, such a helicopter would have very limited range, something along the lines of ninety miles round-trip, if that.

So, whose copter was it and what did its occupants do to the prize bull? The family never found out.

But this weird story gets even weirder. O'Brien first investigated this particular incident thirteen years after it happened. His investigation included doing an exten-sive interview with the bull's owners. The morning after this interview, as he was sitting in his kitchen typ-ing up his notes, O'Brien heard the faint sound of a helicopter approaching. Looking out his window, he was shocked to see an old-fashioned, mustard-yellow whirlybird-type helicopter heading his way. It went by his house at fewer than two hundred feet in altitude, allowing him to see it clearly.

Other members of his family, plus some neighbors, also saw the strange helicopter as it flew past and as it disappeared over the horizon.

The same ultra-rare helicopter, thirteen years later?

"Synchronicity was in effect," O'Brien said. "A truly tricksterish chain of events!"

The Norad Event

The NORAD Event is actually a series of incidents that began at the end of November 1993 and continued until the evening of January 17, 1994.

Over that six-week period, there was an avalanche of reports from people in the SLV documenting multi-colored fireballs, orange orbs, mysterious fires, many mysterious booms, a flurry of Bigfoot sightings and many instances of military activity, and there was also a documented unusual cattle death.

At the height of these events, on January 12, a NORAD official contacted the Rio Grande, Colorado, sheriff's office at three forty p.m. and reported that a significant explosion had occurred about an hour before near the Rock Creek Canyon area. This explosion had been detected by a NORAD satellite scope operator inside Cheyenne Mountain—yet its cause was never determined.

Exactly two hours later Florence, Colorado, resident

Lieutenant Colonel Jimmy Lloyd, a thirty-five-year veteran fighter pilot and self-professed UFO skeptic, reported seeing a group of enormous glowing-green objects in close formation streak over his head and then descend into the San Luis Valley. According to Lloyd, these were not celestial objects such as meteors or any kind of conventional craft or missile. They flew in complete silence as they disappeared from view.

Later on, a source close to NORAD told O'Brien that a newly promoted U.S. Air Force captain in her third trimester of pregnancy had recently been found dead in her garage. The cause: carbon monoxide poisoning. This happened two weeks after the NORAD phone call to the Rio Grande sheriff's office. No death notice was ever carried in the local papers.

Sometime later, O'Brien was told by another NORAD employee that there'd actually been *three* suspicious suicides among NORAD personnel; all had happened around the same time of the so-called NORAD Event.

Meanwhile, a film production company was scheduled to do a film about NORAD. This company had been the first film crew ever allowed aboard a Trident submarine. They had also worked with NASA. Obviously, they were well connected.

But when they mentioned to NORAD that they wanted to do at least part of their film about the NORAD Event, their permission to film inside Cheyenne Mountain was rescinded and the project was killed.

G-O-D or D-O-D?

On August 21, 1994, around midnight, witnesses in Del Norte, Colorado, saw a large group of lights hovering over a nearby mountain.

According to a local newspaper, five of the witnesses said that as the objects hung silently in the night sky, they formed up into a capital letter G, then a circle, then a triangle. It was as if they were spelling out *G-O-D*. . . . Then, one of the objects flew close enough to the witnesses that they could see it was sphere-shaped with blinking red and blue lights. The witnesses watched the mysterious display for almost an hour.

While it's interesting that these people mentioned the mysterious lights forming shapes that suggested they were spelling the word *God*, other witnesses farther east observed the same light show from a different angle and claimed that instead of a *G*, they saw a *D*, spelling out *D-O-D*, which just happens to be the abbreviation for the Department of Defense.

Either way, very strange.

It Gets Personal....

Chris O'Brien's pursuit of the myriad mysteries in the San Luis Valley has led to several rather disturbing incidents for him personally.

In the summer of 1998, he got a call from a man who wanted to pay him a visit and talk about his work. The man then stopped by O'Brien's house. In our interview, O'Brien described him as being "brush cut, and of military bearing."

The visitor identified himself as being from "the Institute of Business and Social Architecture," out of Washington, DC.

After some small talk about O'Brien's investigations, this person told him bluntly, "Maybe you're looking into things that people don't have a need to know."

While O'Brien was reeling a bit from this, the man added, "Perhaps you should question your motivation."

The reason for the meeting seems clear. Either this guy was watching too many movies or someone in the

U.S. government was trying to send O'Brien a not-so-subtle message.

O'Brien experienced an even more chilling episode in the fall of 1998.

It happened on a night that O'Brien was not supposed to be home. His girlfriend was on a business trip and O'Brien and her daughter were supposed to be out of town as well. But plans had changed because his car broke down at a job site, where it sat.

It was the middle of the night when O'Brien heard someone come into his house. Just by the voices, he knew at least three men, maybe four, had come in and were rummaging through his office.

"They were right on the other side of the wall from me," O'Brien said. "They were talking in low voices, but not exactly whispering. It was obvious they thought no one was home."

While O'Brien could hear his file drawers opening and his maps being unrolled, one of the intruders went upstairs, where O'Brien's girlfriend's daughter was sleeping.

"She woke up and was freaked out," O'Brien reported. "She hid under the covers, but caught a glimpse of this guy. She said he was dressed all in black, and wearing something on his head that stuck out from his face."

O'Brien suspects that something was night-vision equipment.

Once the burglars left, O'Brien checked his office and found only that his UFO files had been tampered with—reports, maps and so on. Nothing else had been stolen from the house.

But still this is not the strangest thing O'Brien experienced during his time in the SLV.

That designation goes to the case of the disappearing cargo plane.

O'Brien was outside his home one day and heard the sound of an airplane approaching in the cloudless sky. It was a C-130 cargo plane, a longtime workhorse for the U.S. military and a common sight over any military air base. But there was something different about this particular C-130. It was not the usual dull green or dull gray service color. Instead it was bright and shiny, like it was made of unpainted, polished aluminum or chrome.

Then as O'Brien watched the plane fly directly over Blanca Peak, the plane simply . . . disappeared.

"It blinked out, right before my eyes," O'Brien said.

"There was a flash of light and it was gone. About five seconds later the sound it was making was gone as well. Like someone hit a mute button."

What could it have been? A cloaking device? Some sort of interdimensional transport?

"I don't know," O'Brien said. "I'm still trying to figure that one out."

5

Kirtland Air Force Base, New Mexico

How to Destroy a Human Being

About 120 miles north of Albuquerque, New Mexico, there's a town named Dulce. It's a small place, under three thousand people, and most of it is located within the Jicarilla Apache Reservation.

Nearby is Archuleta Mesa, a peak that rises more than nine thousand feet above sea level. And what's going on inside this mountain appears to make whatever is happening at S4 or in the SLV pale by comparison.

Hidden inside Archuleta Mesa is a multilevel underground complex operated not by the U.S. government but by one (or maybe two) races of horrendously evil space aliens. These ETs are abducting and murdering human beings and eating them. Any leftover organs the aliens use to prolong their own unearthly lives.

But consuming human flesh is just the beginning. These ETs are creating a race of mindless quasi-humanoid creatures by sewing human and animal body parts together. They are also implanting devices inside the brains of ordinary Americans, an alien technology that will turn these people into zombies at the flick of a switch. These zombies will be part of a vast army that will help the aliens take over Earth and ship the rest of us either to the moon or Mars to work as slaves.

As ghastly as all this sounds, the U.S. government is fully aware of what's happening inside Dulce Base. In fact, our government is actually *helping* these alien intruders by providing American scientists to assist in their grisly experiments. Why? Because we earthlings have no other choice. The aliens at Dulce are all-powerful. They can destroy the planet in a snap. They already have millions of humans under mind control, and as they've been here for hundreds if not thousands of years, they actually claim that Earth is *their* planet. So sometime in the 1950s, the U.S. government was given an ultimatum: Either enter into an unholy alliance with the blood-feasting ETs or face a horrifying alternative.

The United States chose the unholy alliance.

* * *

What's actually inside the Dulce complex? A few unauthorized humans have been in there and lived to tell about it.

They report one level is nicknamed "Nightmare Hall." It is here that the ETs do their bizarre crossbreeding experiments. Found within are multilegged creatures best described as half human, half octopus. There are reptiles covered with fur yet possessing human hands. There are various kinds of human lizards plus fish, birds and mice so grotesquely mutated they are unrecognizable.

The Dulce complex is said to be nearly impenetrable. The security forces carry "flashguns," weapons that work effectively against both humans and aliens. Those humans forced to work inside the complex have their own fashion: They wear jumpsuits with the Dulce symbol on the front upper-left side. And whenever U.S. government higher-ups secretly visit the site, they carry ID cards containing the Great Seal of the United States, the all-seeing-eye symbol that links all those "Harvard-types" with the Illuminati's New World Order.

Once inside Dulce, each authorized visitor is stripped naked, weighed and then given an off-white uniform. Each doorway in the complex has a weight scale in front of it. If you weigh more standing at any given door than you did when you came in, security is called and you'll have a flashgun stuck in your face.

But *who* are these nasty aliens?

By many reports, there are actually *two* types of ETs living inside Dulce. One species is the familiar big-headed Grays. The second is a race of lizard-like aliens some simply call the Reptoids. Though these Grays and Reptoids are in league with each other, it's a strained relationship; they are uneasy allies at best. There are said to be as many as eighteen thousand Grays and Reptoids combined living inside Dulce Base, doing their dastardly things and preparing to enslave Earth.

And again, there's not much humanity can do about it.

Not that we haven't tried. In 1979, a small army of U.S. Special Forces attempted to oust the aliens from the mesa. But the attack failed, at the cost of many human lives.

Since then, the U.S. military has been forced to keep out of the affairs at Dulce.

But what about our elected leaders? What have they known about this place?

Too much, as it turns out. History says Dwight D. Eisenhower was the first president to be made aware of this silent invasion. He was forced to sign an alliance with the aliens back in the fifties. Another president, upset about the intimate help some of America's intelligence agencies were giving to the alien intruders, told them to cease playing footsie with the ETs—or he was going to reveal the whole sordid mess to the world.

That president was John F. Kennedy.

And we know what happened to him. . . .

What are the aliens implanting in us? Brain transceivers as it turns out. Inserted into the nose and up into the cerebrum, these transceivers allow the aliens to transmit telepathic messages and manipulate the minds of their unfortunate hosts.

These brain transceivers are used with full knowledge of the U.S. government as well as the governments of Russia and, inexplicably, Sweden. If this arrangement suggests that the United States and Russia haven't exactly been enemies over the years, apparently that's correct.

In fact, secret bases built and run by a joint American-Russian-ET cabal have existed for years on the dark side of the moon and on Mars. Once the aliens decide the time is right, the human race will be fully subjugated and shipped off to either of these unheavenly bodies to work as slave laborers.

But why do the aliens need human slaves, androids and bizarrely composite human-animals? As disposable beings to work on their dangerous plutonium rocket and saucer experiments, of course.

So what can we, the human race, do about all this?

Nothing, apparently.

But as individuals, there *is* one element we should all be aware of.

Apparently some of the aliens, particularly the Grays, have become angry with humans for eating chocolate.

According to some reports, if you eat chocolate, the aliens can't use your bodily fluids.

So, if we all eat a lot of Hershey's bars, then maybe we can . . .

Okay . . . let's stop the nonsense right here.

Not that there isn't more when it comes to Dulce Base. Just the opposite; there's *plenty* more: Books, DVDs, conferences, lectures, websites, newsletters, radio shows, TV shows—it seems there's an endless supply of stories, theories, rumors, conjecture and just

plain foolishness when it comes to Dulce and the eighteen thousand flesh-eating aliens that live within it. In
fact, all of the above was provided by the cornucopia of
silliness that can be found via a simple Google search
of "Dulce Base."

The question is, how did it all start?

As it turns out, the real story is almost as bizarre.

With help from conversations with and the writings
of two outstanding UFO authors, Jerome Clark (the
multi-volume *UFO Encyclopedia*) and Nick Redfern
(*Keep Out*), we can try to piece together what happened.

The Concerned Citizen

The story of Dulce Base is actually the story of an unfortunate man named Paul Bennewitz.

Bennewitz was a physicist who ran a company called
Thunder Scientific Labs. This company was located
next to Kirtland Air Force Base, which is just outside
Albuquerque, New Mexico.

Kirtland AFB is an enormous facility today, encompassing a number of missions for the Air Force, including airborne laser weapons testing and nuclear weapons
storage. Years ago, there was a place within the massive

fifty-two-thousand-acre site called the Manzano Storage Facility. This was where the Air Force kept many of its nuclear weapons at the time.

In the late 1970s, Dr. Bennewitz began monitoring strange electromagnetic signals he believed were coming from the Manzano facility. At the same time, he'd been sighting strange objects flying over this same part of Kirtland, specifically above entrances to secret tunnels where the base's nuclear weapons were stored.

Bennewitz put these two elements together and believed he'd come upon something rather startling: that the objects he spotted flying over the nuke storage facility were UFOs and the signals he'd been intercepting were coming from them.

Bennewitz first reported all this to a Tucson-based UFO research group. However, the group thought he was delusional, so they dismissed him out of hand.

Bennewitz did not give up, though. In late 1980, he went directly to the Air Force with his suspicions, saying he was convinced that the nuclear weapons being stored at Kirtland were under some kind of threat by whoever or whatever was piloting the UFOs.

Surprisingly, the Air Force gave him a sympathetic

ear. Two members of the Air Force Office of Special Investigations (AFOSI) interviewed Bennewitz at his Albuquerque home. Encouraged that the military was listening to him, Bennewitz played the investigators several recordings that seemed to indicate high electromagnetic signals coming from the Manzano area. Bennewitz also showed them photographs of the aerial objects he'd spotted over Albuquerque and reiterated his theory: that the UFOs in the pictures were producing the weird pulses over the nuclear storage area.

A few weeks went by. Then, on November 10, 1980, Bennewitz was asked to come to Kirtland itself and brief a larger group of Air Force officers and scientists.

At that point, Bennewitz must have felt on top of the world.

Right Place, Wrong Info

Bennewitz had indeed uncovered some startling information—but it had nothing to do with UFOs.

What he'd stumbled upon was a highly classified National Security Agency communication system as well as the Air Force's capability of secretly tracking the movements of Russian spy satellites. And those strange flying objects Bennewitz had observed over the base? Most likely they were early test flights of stealth aircraft prototypes.

So, the military *was* interested in Bennewitz—not for his beliefs on UFOs and alien messages but because he'd actually managed to uncover some authentic secret programs. The military's fear was that Bennewitz would speak openly about what he'd found, claiming it was about UFOs, when he'd actually be revealing invaluable information to the Russians.

The Kirtland AFOSI decided this could not happen—even if it meant Bennewitz would have to suffer as a result.

So they set out to drive Dr. Bennewitz insane.

The Government's Evil Plan

Around this time, members of the AFOSI reportedly burglarized Bennewitz's home. Breaking into his computer, the culprits were able to read his files and learn that the physicist had formed some beliefs that went beyond what was happening over Kirtland. For instance, Bennewitz was closely following reports of cattle mutilations in the area and speculating they were linked to the UFO activity he believed he'd uncovered.

Moreover, he was theorizing that the same aliens might also be abducting ordinary citizens as well. Most telling, however, was that the physicist had become convinced this ET activity was somehow connected to Dulce, New Mexico.

This was perfect for what the Air Force was planning. They decided the best way to maneuver Bennewitz away from Kirtland AFB was to keep him pointed at Dulce and somehow convince him that his unorthodox theories were true.

To this end, the Air Force began providing Bennewitz with seemingly official information on Dulce. This included faked "official" documents containing tales of extraterrestrial activity going on inside the mesa.

The Air Force overloaded Bennewitz with so much of this disinformation that the physicist was certain he was on the right track. And why wouldn't he be? He believed what his government was telling him.

But soon enough, the reality of that unreality would get the best of him.

Armed, Then Committed . . .

Bennewitz became especially involved with the mystery of animal mutilations. He'd met a woman who claimed she and her young son had witnessed a calf being cut up by aliens. When the aliens realized they were being watched, the woman said they performed a procedure on her and her son that caused them to suffer from confusion and lose their memory.

Fascinated by her account, Bennewitz arranged for both the woman and her son to be hypnotized, hoping more of the real story would come out. Indeed, under hypnosis the woman told a graphic tale of being abducted and taken aboard the alien craft, where she saw animal and human body parts stored throughout the UFO.

But Bennewitz grew increasingly paranoid as these hypnosis sessions progressed. He began to suspect the hypnotist was a CIA agent. Finally, he made the hypnotist stop the sessions and forbade him to talk to the woman again. The hypnotist later reported that by this point Bennewitz had armed himself as protection against the aliens.

Becoming even more irrational, Bennewitz began preaching his theories to anyone who would listen. This was around the same time that he was first institutionalized, one of three such occasions when he was committed. His crazy ideas about Dulce began to spread throughout America's substantial UFO community, and, as is sometimes the case within that community, the stories grew in complexity and size—and, with each retelling, became more outrageous and bizarre.

Finally, in 1989, one of the people who'd been funneling the disinformation to Bennewitz admitted his role in the affair. (Surprisingly, it was a fellow ufologist secretly recruited by the AFOSI.) This man confirmed that the government's goal all along was to push Bennewitz into a mental breakdown by feeding him false information about aliens.

An Air Force sergeant assigned to the AFOSI named Richard Doty was also outed for taking part in the scheme. He and the ufologist turncoat damaged Bennewitz so severely that even when Doty contacted Bennewitz years later to tell him he'd been an unwilling participant in the government-inspired hoax, Bennewitz refused to believe him.

Was all this necessary? Why didn't the Air Force simply sit down with Dr. Bennewitz and explain the real situation to him and have him sign a national security document promising never to speak about what he'd stumbled upon? Thousands of ordinary citizens and military personnel have signed similar documents over the years. Why wasn't Dr. Bennewitz afforded the same opportunity? From the beginning, all he wanted to do was help his government. But in the end, that same government destroyed him.

"It was cruel," said Jerome Clark of the entire episode. "And it was done by someone who was in a position of power and could do it. So they did."

But Clark also questions why the U.S. Air Force and others had such an extraordinary interest in this single ufologist in particular and in the UFO phenomenon overall. Indeed, our beer-drinking Spook friend told us that, as described, the whole Bennewitz affair most certainly went higher than just Kirtland's local AFOSI.

"Something like this would have to have at least several agencies involved," he said.

So, why would the U.S. government go through all this trouble and expense if there were no real UFO secrets to protect?

The answer to that question remains elusive.

And though Paul Bennewitz died in June 2003, the nonsense about Dulce Base lives on.

6

Tonopah and the Tale of Two Cities

Flying with Ghosts

On the night of May 19, 1900, a prospector named Jim Butler woke up to discover his burro was missing.

Butler was camped out in a rugged, mountainous part of south central Nevada, picking through rocks and looking for fortune. About forty years earlier, the Comstock Lode, one of the largest silver finds ever, had been discovered about two hundred miles to the northwest. But the lode had been stripped clean by 1889, bringing about what some people believed was the end of all mining in Nevada.

A few die-hards like Butler kept at it, though, searching other parts of the state and hoping to find their pot of gold—or silver. Such work required a pack animal to carry around the tools of the trade, so hanging on to one's burro was important.

The next morning Butler set off in search of the animal. After some hiking and climbing, he finally found it sleeping under a small ledge.

Determined to get the burro's attention, Butler picked up a rock to throw at it—but the rock seemed unusually heavy. Examining it closer, Butler realized this was no ordinary rock. It was thick with both gold and silver. Though Butler didn't know it at the time, thanks to his wayward donkey he'd just uncovered the *second-*richest silver strike in Nevada history.

Butler and others went about claiming stakes in the area, kicking off a spectacular twenty-year mining boom that historians say breathed life into what was at the time a dying state.

Butler himself named the town that sprang up around all this. He called it Tonopah, a Shoshone Indian word meaning "hidden spring."

These days Tonopah is a junction town, a place where U.S. Routes 6 and 95 cross, putting it about midway between Las Vegas and Reno, Nevada.

Because the mining business dried up so long ago, the town (population now about 2,500) appears to be little more than a place to gas up and maybe spend the night before moving on to the gambling mecca of your choice. (The iconic Clown Motel is considered one of the top places to stay; one reviewer on tripadvisor.com called it "affordable, funky, with the best shower in six states.")

Tourist brochures also cite Tonopah as a top stargazing destination, which will be important later on. Plus, the ghost town of Goldfield is just twenty minutes down the road.

But sleepy little Tonopah is not what it appears. Strange things have happened there in more recent years.

Very strange and very secret things.

The Most Mysterious Place Ever?

Thirty miles southeast of Tonopah there's a highly classified military base few people have ever heard of.

Known as the Tonopah Test Range, or TTR, this base is so secret that if one of the ground vehicles used there breaks down, the person hired to fix it must have the highest government security clearance possible. In fact, driving from the base into the town of Tonopah without special permission is prohibited. The TTR's on-base security is among the most stringent of any U.S. installation around the world.

For years, personnel assigned here lived like vampires, staying hidden during the day and venturing out only at night. There was a good reason for this. Hidden at the base for nearly a decade was a flying machine of incredible capabilities. A machine so hush-hush it had to be concealed inside a slew of secretly built hangars whose doors, by the strictest orders, could not be opened until one hour after sunset and had to be closed one hour before sunrise. A machine so remarkable that, by presidential order, it only flew at night.

Had any civilian aircraft strayed too close to the TTR while this machine was airborne, there was a good chance they'd have been chased off by patrolling Air Force fighters. Though it wouldn't have made much difference. Because to the pilot of any accidental intruder flying at night, Tonopah's secret aircraft would have been invisible.

At the same time, though, if some plucky soul climbed one of the high mountains that surround the TTR and looked out on the base through night-vision goggles, they might have caught a lucky glimpse of this mysterious flying machine.

How could it be seen and yet not seen?

This seemingly otherworldly aircraft was not a reengineered UFO or something built with the help of aliens, though it *did* have capabilities many people have seen in UFOs: Here one moment, gone the next. A very bizarre, unearthly shape. The aforementioned ability to disappear.

Officially this flying machine was called the F-117 Nighthawk.

To most of us, it was better known as the Stealth Fighter.

An Early Triumph

The Nighthawk was basically an F-15 Eagle fighter with a different body. The secret was that body wore highly classified radar-absorbing paint, had no right angles and had many of its heat sources shielded or deflected—plus a number of other things that reduced its radar signature to the size of a pea.

Back in the early 1980s, during one of the plane's first trial flights, a radar station was set up in the middle of the desert as a test of sorts. The idea was, if the Stealth Fighter were a "ghost" then it would not show up on the radar as it approached. At the appointed time, though, the plane's designers were crushed when a blip did pop up on their screen. But a few seconds later, that disappointment turned to triumph when they realized what they were looking at was the radar signature not of the Stealth but of its chase plane.

So, the Stealth itself was indeed "invisible," as would be every major U.S. warplane to follow.

Reds in the Desert

The F-117 secretly flew out of the TTR for almost ten years. Yet the place was highly mysterious even before the Stealth's arrival.

Again, had someone managed to climb up one of the surrounding mountains in the 1970s or early 1980s, day or night, and done so at just the right time, they would have seen a very different aircraft flying above Tonopah.

They might have been confused at first, though, as if what they were seeing was something from an alternate universe. Because in those days, the skies over the TTR were filled with Soviet MiGs.

What was the Russian Air Force doing at Tonopah?

In the days before the Stealth Fighters arrived, the U.S. Air Force had managed to secure a number of

Russian fighter aircraft, MiG-17s and MiG-23s mostly, and bring them to America. Just how they did this is still a secret, though at least a few of the foreign airplanes came by way of defectors—lured by a reward of a million dollars or more—flying in from North Korea or other Soviet client states.

These Russian airplanes, laid out in full markings and wearing Soviet camouflage paint, became part of a MiG air-combat training program. Flown by highly trained American pilots, the Russian planes took to the air as adversary aircraft, giving other U.S. fighter pilots a chance to practice dogfighting techniques against the top-line aircraft of their main Cold War rivals.

Fans of the movie *Top Gun* will remember Tom Cruise and his Navy buds also flew against adversary planes—but those were U.S.-made aircraft painted to look like Soviet aircraft. Up in Tonopah, away from Hollywood's klieg lights, the U.S. Air Force was doing it for real.

The New, Expensive Occupant

When the Soviet threat began to change somewhat in the early eighties, the Russian jets moved out and the Stealth program moved in.

In 1982, the Tonopah base was modernized to accommodate the invisible jets. An enormous complex, including seventy-two specially built hangars, was constructed to house and hide the remarkable F-117s in the daytime.

But strict procedures were still needed to keep the secret planes secret whenever they were flying at night. To that end, before each mission, there would be a mass briefing of all Stealth pilots. Again, Tonopah's hangar doors could not be opened until one hour after sunset. This meant the first takeoff could not be made until about seven p.m. in winter or nine p.m. in the summer. At the end of the night, the planes had to be in their hangars and the doors closed one hour before daylight.

Moreover, a surrogate aircraft was needed to provide cover for the Stealths. This role fell to the A-7 Corsair, a Vietnam-era attack plane that was approaching the end of its service life. Operating from Tonopah throughout the F-117 program, Corsairs were intentionally left outside their hangars in the daytime so that Russian satellites looking down on the TTR would detect nothing but the antiquated airplanes during their photo passes. Whenever the Stealths flew off range (mostly over other parts of Nevada and California), conversations between the F-117 pilots and the Tonopah control tower were doctored to make it seem to anyone listening in that they were all about flying and controlling an A-7. Each F-117 aircraft also carried a radar transponder that mimicked that of an A-7.

All of this was necessary to keep the existence of the Stealths secret.

But it was a very expensive cover story.

Where Are the Little Green Men?

So what does all this have to do with UFOs?

Very little, as it turns out.

Though the TTR does raise a few eyebrows among the conspiracy community simply because classified aircraft are known to fly from there, it's not really considered a hot spot for UFOs.

Proof of this comes from the previously mentioned tourist brochures: In fact, Tonopah claims to be the "stargazing capital" of the United States. In other words, unlike Groom Lake and other secret bases around the world, people come here to look at the stars and not for UFOs, which means there can't be many of them flying around, or any at all.

But . . . this is where it gets weird.

Because, as it turns out, people *do* see UFOs over Tonopah. Lots of them.

Not Tonopah, Nevada, though, but Tonopah, *Arizona*.

Same Name, Different Place

Located five hundred miles southeast of Nevada's Tonopah, Arizona's Tonopah can be found along Interstate 10 about an hour west of Phoenix.

This Tonopah is so tiny, though, it barely qualifies as a town. At last count, fewer than one hundred people lived there. It's located in the desert and sits on top of a huge underground aquifer that dispenses well water as hot as 120 degrees.

But this place, and the surrounding area, is brimming with something else: UFO activity.

Tonopah, Arizona, has no secret air base close by, but it does have something UFOs are known to take an interest in. This Tonopah is near the Palo Verde Nuclear Generating Station, the largest nuclear power plant in the United States.

And not only have there been many UFO sightings here, but there has also been at least one CE3 (as in, close encounter of the third kind) event reported in the area—that is, an actual meeting between humans and the occupants of a UFO.

Many of these UFO sightings have taken place along the notorious Interstate 10. The road is typical of Arizona desert highways. For the most part it's straight as an arrow with only an occasional rise or hill. The desert is vast on either side, with hundreds of square miles of sagebrush and sand, the only break being a sign reminding you just how far away you are from anywhere else.

Isolated, flat and lonely, it's the perfect place to encounter a UFO.

Courtesy of MUFON.com, NUFORC.com, www .forteanswest.com and wwwufospottings.com, the incidents below are just a small sampling of some of the strangest UFO reports from the "other" Tonopah.

UFO ROAD RAGE

On July 7, 1989, a man was traveling east on Interstate 10. It was about one thirty a.m. and he was going approximately seventy miles per hour. Suddenly he became aware of something in the fast lane traveling alongside him. It startled him so much he nearly went off the road. He thought it was a truck cab driving with no headlights. But he could see no mud flaps or reflector lights.

The man quickly realized that whatever this was, it had no discernible shape. Nor did it have any wheels.

Rather it was *flying* alongside him, just a few feet off the ground.

After a few terrifying seconds the object accelerated, pulled ahead of his car and began to climb. The astonished driver found himself leaning forward in his seat and looking straight up through his windshield. While this helped him get a better look at the object, he still had no idea what it was—other than it was not a plane or helicopter or ultralight aircraft.

The object accelerated further and finally flew off into the night. The driver skidded to a stop and jumped out of his car just in time to see it go up and over a nearby mountaintop.

THE JET FIGHTER AND THE UFO

On the night of March 14, 1997, a man was driving west on Interstate 10 when he noticed an unusually bright light directly in front of him. It was brighter than any star or planet and it was moving toward him at high speed, revealing its enormous size. When the object suddenly changed direction, the witness realized that within it was a cluster of three separate bright objects. Even more astonishing, a jet fighter was circling these objects, as if examining them.

Asked to make a comparison, the witness said if the three bright lights were the size of golf balls then the jet aircraft would have been about the size of the head of a tack. In other words, the three objects were huge.

Suddenly, the bright lights disappeared and the only thing left in the sky was the jet fighter. The witness watched the jet make one more long arc around the area and then fly off to the south. The witness checked the time and realized he'd been watching this drama for at least twenty minutes. He later said it was only after the sky was empty that the reality of what had happened finally hit him.

The witness later told UFO investigators that the sighting had been the most thought-provoking and unexplainable incident of his life.

THE PERFECT V

On the night of November 16, 2011, a witness was walking outside his home near Tonopah, Arizona, when he noticed strange lights in the sky off to the northwest. There were five of them and they were brighter than any star or planet. Glowing all white, they were evenly spaced about forty degrees above the horizon. As the witness watched, the lights suddenly blinked off but just as quickly came back on again in sequence, from top to bottom. The lights went off a second time, but when they blinked back on again, an identical line of lights had appeared next to the first.

The witness was astonished to see both lines move to form a perfect V in the sky.

The entire event lasted less than a minute before blinking out for good. The witness later stated that he had graduate degrees in geology and geography and, while he was definitely not a UFO enthusiast, he had come forward simply because he wanted to tell someone of his strange experience.

CE3 IN THE DESERT

Quite possibly the strangest UFO incident in the Tonopah, Arizona, area occurred on July 1, 2009. Around ten thirty p.m., four young people, two males and two females, driving near the Palo Verde nuclear plant spotted a triangle-shaped UFO.

When the UFO slipped behind a mountain, the witnesses pursued it. Coming around the mountain, they saw a flashing light.

The witnesses drove to within one thousand feet of the light and realized it was coming from the object they'd just seen. It was close to the ground and huge. At this point, the two males approached the craft. One recorded the events on video.

The object had numerous windows, through which occupants were seen. Some were small with big heads. Others were tall and thin. The craft landed and the

BEYOND AREA 51

pair drew closer. A doorway opened and two aliens appeared. The terrified witnesses tried to run but could not move.

The aliens, one short and one very tall, came within twenty feet and communicated telepathically. They explained their mission was to prevent us from destroying the Earth.

Their next message was chilling: A horrible event was going to happen in the Middle East, which had to be stopped.

After saying they would meet again, the aliens returned to their craft. It vanished in an instant.

The shocked pair made it to their vehicle. The witnesses stated that as they drove away, three military helicopters descended on them. Moments later, two military Humvees appeared and, via a loudspeaker, ordered them to immediately pull over.

They were asked what they'd seen, for how long, and if they'd had any contact with the craft. The military confiscated the witnesses' camera and took one of the young men with them for questioning. He was later dropped off at a local diner, where his friends picked him up.

Back in Nevada

After years of secret operations at the TTR and following their deployment to the Persian Gulf in 1992, the Tonopah Stealths were eventually sent to Holloman Air Force Base in New Mexico.

This redeployment began the Stealths' road to decommissioning. While it's hard to believe that such a magical airplane would ever become obsolete, America's newer fighters, like the F-22 Raptor and the F-35 Lightning II, not only inherited many of the F-117's secrets, but they also improved on them. In April 2008, the Stealth Fighter fleet was quietly deactivated and put in mothballs.

But it had cost millions to house the F-117s at Tonopah and millions more to operate the base, support the personnel and keep it all secret. Keeping secrets is expensive, and in the world of black ops, the government tends to stay on familiar ground. So, what took the Stealths' place?

As one former fighter pilot who used to fly all over the Nevada Weapons Range told us, "If something moved out of Tonopah, that only means that something else even more secret moved in."

The question is, what?

The latest official line, courtesy of a Department of Energy website, is that Tonopah is currently being used for "nuclear weapons stockpile reliability testing, research and development of fusing and firing systems and testing nuclear weapon delivery systems. These capabilities allow Tonopah to support directed stockpile work, the ability to perform surveillance testing on nuclear bombs and compatibility with the Air Force bombers and fighters. Stockpile surveillance is vital."

But when we ran this explanation by someone who works with the U.S. intelligence community, their blunt reply was: "That's bullshit—whatever they are really doing there, that's a cover story."

Leaving Clues in Plain Sight

Foreign-made weapons systems are scattered all over the Tonopah Test Range these days.

A number of websites featuring satellite imagery show Russian-made surface-to-air missile systems and so-called MAZ artillery trucks dotting the TTR's landscape. Other photos show large radar dishes, Scud missile launchers and even Russian military helicopters at the TTR as well. Like the Soviet fighters that once flew the TTR's skies, these Russian weapons came into U.S. possession—we can only assume—via backroom deals with so-called un-known suppliers.

And we have to assume, as these images are readily available on the Internet, these foreign-made weapons must surely show up on spy satellites belonging to rival superpowers.

But maybe that was the plan all along.

Remember the A-7 Corsairs that were operating out of Tonopah when the Stealths were flying?

They were left out of their hangars in the daytime to fool Russian satellites passing over into thinking that nothing too grand was happening at the TTR. Now, thirty years have gone by and the Stealths have been deactivated, yet we can still see interesting things scattered across the Tonopah range, essentially in full view.

But past experience tells us that what can be seen in the daytime at Tonopah is most likely part of a cover story. And like those vampire personnel of the 1980s, whatever is happening at the TTR these days probably only comes out at night.

As confirmed by a friend in the intelligence community, the base's extra long, twelve-thousand-foot runway is still active, along with its navigation aids. The strict rules of behavior for all personnel assigned to the TTR are still in place. Everyone on the base must be a U.S. citizen and have no felonies on their record. No private vehicles are allowed anywhere on the base. No recording devices, no cameras, no laptops, no hand-held devices.

So, the base is still set up as if something supersecret is being done there.

What could that be?

We know that if there was ever a place to teach U.S. pilots to fly unusual aircraft, or to have those same pilots fly *against* unusual aircraft, Tonopah is probably that place. But beyond that would be mere speculation.

However, the bigger question might be, why do UFOs avoid this place but are very active around a place of the same name, just in a different state? As we will see, nearly every secret base around the world is the site of at least *some* UFO activity, real or imagined.

Why is Tonopah, Nevada, different? Why does it have zero UFO activity?

Whatever secret weapons the United States is testing there, could the weapons be so unusual that even UFOs stay away?

Maybe the only way to know for sure is to catch a glimpse of what's inside those mysterious hangars when they open up exactly one hour after sunset—but maybe not even then.

As our Spook friend told us, "People in the intelligence community talk about Area 51 all the time. But no one ever talks about Tonopah."

7

Homestead
Air Force Base

Unlikely Friends

Jackie Gleason was born February 26, 1916, in Brooklyn, New York.

After a rough childhood, during which his only brother died and his father walked out on the family, young Gleason joined a street gang at age fourteen and began hustling pool. Though he would eventually drop out of high school, a part in a class play went well enough for him to get an MC job at a local theater. He then went on to work as a stunt diver, a daredevil driver, a disc jockey and a carnival barker, among other things.

By 1935, Gleason was a professional comedian,

working nightclubs in New York City. Within five years he was in the movies. By this time, to say the rotund entertainer was gregarious would be an understatement. Gleason threw such loud and raucous parties in his LA hotel suite he had to soundproof it at the insistence of the management.

In 1954, Gleason made it onto television, hosting a variety show. He also enjoyed a musical career, recording more than thirty-five albums, and would eventually win a Tony for acting on Broadway.

But Gleason was probably best known for his hit TV show *The Honeymooners,* in which he played big-dreaming bus driver Ralph Kramden.

The show was so popular, an eight-foot-tall bronze statue of Gleason's Ralph Kramden character still stands today outside the Port Authority Bus Terminal in New York City.

Richard Nixon was born on January 9, 1913, in Yorba Linda, California.

The product of a strict father and a Quaker mother, Nixon saw two siblings die at a young age, both of tuberculosis. After high school, Nixon attended Whittier College and then went on to Duke University School of Law, graduating in 1937.

After serving in the U.S. Navy during World War II, Nixon was elected to the U.S. House of Representatives in 1946 and then to the U.S. Senate in 1950. Known as an ardent anti-Communist, he became vice president under Dwight D. Eisenhower in 1952 and then again four years later. Though Nixon barely lost his own bid for the presidency to John F. Kennedy in 1960, he ran again in 1968 and won.

Nixon dramatically escalated the Vietnam War before finally ending America's involvement in 1973. Though still touted as a virulent anti-Communist, he made a historic visit to Red China in 1972 and improved relations with the Soviet Union as well.

Nixon was also in office when men first walked on the moon. Speaking by phone to astronauts Neil Armstrong and Buzz Aldrin while they were on the lunar surface, Nixon enthusiastically called the conversation the "most historic phone call ever made from the White House." However, when NASA proposed plans to establish a permanent base on the moon, as well as a trip to Mars by 1981, Nixon vetoed both, essentially killing America's manned exploration of outer space.

Nixon's second term as president was rife with troubles, not the least of which was his involvement in the Watergate scandal. On August 9, 1974, he resigned the presidency in disgrace.

Often portrayed as unshaven and sweaty, Nixon was also overly secretive and awkward. He was so straitlaced he wore a suit coat and tie even when he was home alone. Historians have called him the most peculiar of all the presidents, someone who assumed the worst in people and brought out the worst in them.

So it's strange, then, that these two men—Jackie Gleason, the heavy-drinking, chain-smoking, boisterous wise guy from Brooklyn, and Richard Nixon, the dark, self-conscious, corrupt politician from California—would be good friends. Yet they were.

But even stranger is the adventure many claim they shared at a place called Homestead Air Force Base.

A Historic but Stormy Place

Located approximately twenty-five miles south of Miami, Florida, Homestead Air Force Base was established in 1942 as a major stopover point for U.S. combat aircraft going to the Caribbean and North Africa.

In November 1955, the U.S. Air Force's 379th Bombardment Wing was assigned to Homestead and consisted of four squadrons of nuclear-armed bombers. In an ironic twist, Homestead's 379th Bombardment

Wing inherited the colors of the World War II–era Army Air Force 379th Bombardment Group whose pilots and crews had encountered many "foo fighters" over Europe in World War II.

In the ensuing years, Homestead became home to a number of USAF units, many of them nuclear-armed. This was crucial during the Cuban missile crisis in October 1962 due to the base's proximity to the Communist-controlled island, just two hundred miles to the south, or about twenty minutes' flying time in a modern jet fighter.

At present, Homestead is home to number of U.S. military units, including those belonging to the Air Force Reserves, the Florida National Guard and the U.S. Coast Guard. Because of its location, though, Homestead has been a frequent victim of hurricanes. An especially violent one hit the base in 1945. But in 1992, Hurricane Andrew, one of the most powerful storms to ever hit the United States, virtually wiped Homestead off the map.

But back in the early 1970s, Homestead was a sprawling installation operated by the USAF's Tactical Air Command. It had several extra long runways, dozens of hangars and support buildings, and many areas considered classified and off-limits to individuals without high security clearance.

This is what Homestead looked like when the following story is said to have taken place.

A Trip at Midnight

Gleason and Nixon shared a love of golf. They would play together when both were in Florida, which was frequently. Gleason spent the latter part of his career living in the Miami area, and Nixon had a home in nearby Key Biscayne.

As the story goes, during one such golf outing in early 1974, the subject of UFOs came up.

Gleason had long been interested in UFOs. Not only did he possess nearly two thousand books having to do with UFOs and other paranormal subjects, but the comedian's home in Peekskill, New York, was actually shaped like a flying saucer. By Gleason's orders, everything inside the house was also saucer-shaped, including the furniture. Gleason called this house the "Mothership" and its garage, which also looked like a flying saucer, the "Scout Ship." (Ironically, Gleason had a fear of flying and mostly traveled by train.)

Gleason's biographer, William Henry III, confirmed in his book *The Great One: The Life and Legend of Jackie Gleason* that Gleason had a lifelong fascination with the paranormal. And as Gleason was an insomniac, to pass the time at night he would read his UFO books.

But Gleason kept his fascination with UFOs more or less to himself, knowing if any celebrity gossip magazines knew the extent of his interest, he'd be labeled as a UFO nut.

And while Nixon most likely knew of Gleason's curiosity about UFOs, they'd never had a deep discussion about the subject, probably because Nixon was usually surrounded by presidential aides and Secret Service men.

But that day on the golf course, they talked about the phenomenon at length.

Later that night, when Gleason was back at his Florida home, Nixon suddenly appeared at his door. It was around midnight, and the chief executive was alone, with no Secret Service agents in sight. (This was not so unusual; Nixon was famous for giving his Secret Service detail the slip.)

Still, Gleason was shocked to see Nixon. The president told Gleason he wanted to show him something. With that, they drove to nearby Homestead Air Force Base.

By at least one recounting, when they arrived at the base's main gate they were challenged by an MP, who initially stopped the car from going any farther. On looking inside and seeing Nixon, though, the astonished MP turned pale, then saluted and let the car pass.

The president and the comedian drove to the far end of the facility, finally stopping at a well-guarded building. Aware by now that Nixon was on the base, the security police simply parted the way for them.

Nixon and Gleason entered the building to find a large number of laboratories and such, but nothing too unusual. But then Nixon brought Gleason to one room where he showed the comedian what he described as the remains of a flying saucer. Gleason later told friends that he was sure this was all some kind of joke, because he and Nixon had spoken about UFOs on the golf course earlier that day. But Gleason quickly realized Nixon was serious.

Nixon next brought Gleason to an inner chamber where a half-dozen freezers were located. They had glass tops, allowing Gleason to look down into them.

What the comedian saw, he thought at first, were the mangled remains of some children—perhaps victims of some kind of accident.

But then Gleason took a closer look and realized whatever he was looking at appeared extremely aged, with large oval eyes and gray skin—and certainly not human.

Gleason was so shocked that he and Nixon left the classified facility almost immediately.

No Stranger to Aliens

This was not the first time that Homestead Air Force Base had been mentioned in connection with UFOs—the place has an interesting past when it comes to extraterrestrial craft and their occupants.

There is a story, famous in UFO folklore but with a surprising amount of documentation, that claims on the weekend of February 20–21, 1954, then President Dwight D. Eisenhower went missing. The official story was he'd gone to a dentist's office under the guise of having chipped his tooth earlier that day. But many UFO researchers claim he was actually meeting a delegation of extraterrestrials to talk about how humans and an alien race could coexist on Earth.

True, this story is far-fetched and probably belongs in the Dulce file. Besides, in its many retellings, some claim it actually happened at Edwards Air Force Base in California.

But other UFO researchers insist this meeting

between Eisenhower and the aliens took place at Homestead Air Force Base. True or not, it seems that Eisenhower was indeed unaccounted for during the night in question.

To the Moon, Beverly

Jackie Gleason never spoke openly about his strange visit to Homestead Air Force Base. In fact, most of the information about the incident comes from Gleason's second wife, Beverly.

She spoke to *Esquire* magazine after she and Gleason separated later in 1974. At the time she was considering writing a book about their marriage.

The UFO story came up in that interview and she confirmed Gleason had been out late one night and when he returned home he had told her that he'd been to Homestead Air Force Base with President Nixon— and that he had seen some dead alien bodies, an event that traumatized him for the next several weeks. Beverly Gleason also confirmed that her husband and Nixon were frequently in touch, and whenever they wanted to play golf together someone on Nixon's staff would set it up.

But the ex–Mrs. Gleason soon knew she'd made a mistake by talking to *Esquire* about the Homestead incident, because not too long afterward Jackie called to tell her he didn't appreciate her giving the interview—and was especially upset that the UFO story had gone public.

Some reports say this was the incident that finally led to their divorce.

"Jackie's Right. . . ."

Gleason died in 1987, Nixon in 1994. With his passing, Nixon was remembered by most to be the only U.S. president to resign the office, his legacy one of deception and dishonor.

Gleason on the other hand was remembered not just for his TV celebrity and *The Honeymooners* but also for making a comeback in movies, playing a good-old-boy sheriff named Buford T. Justice in the Burt Reynolds *Smokey and the Bandit* films and then going on to star with the likes of Tom Hanks and Richard Pryor in two highly successful big-screen comedies.

After he died, Gleason's third wife donated his huge collection of UFO books to the University of

Miami library. He left nothing in his estate to indicate whether the story of his trip to Homestead Air Force Base was true or not—no evidence, no notes to friends or family.

This was no surprise as, again, there were very few people in or out of show business whom Gleason trusted to talk to about UFOs. But according to published reports, one person he did confide in was Bob Considine, a famous and highly influential columnist for the Hearst newspaper chain.

The two men would frequently dine at a famous New York City restaurant owned by Jackie's friend Toots Shor and argue about the existence of UFOs. During these lively debates, Gleason tried very hard to convince Considine, a nonbeliever, that UFOs were real, telling him that pilots on both sides—the Allies and the Axis—saw UFOs during World War II. Gleason also was said to have claimed that as many as four presidents had told him that they knew UFOs existed.

But Considine would not budge . . . until one day, when an Air Force officer happened to be in Shor's restaurant and overheard the two men arguing.

This man was no ordinary officer. He was General Emmett "Rosie" O'Donnell, commander in chief of U.S. Pacific Air Forces from 1959 to 1963 and leader of the first large-scale attack on Tokyo during World

War II. O'Donnell was a war hero and someone who was a straightforward, no-nonsense type of guy.

That day, after hearing Gleason and Considine, O'Donnell walked over to their table and politely interrupted them, then said two words to Considine: "Jackie's right."

8

The Navy's Area 51

The Weirdest Edge of the Triangle

More than fifty ships and a thousand people have vanished inside the Bermuda Triangle over the past few hundred years. At least twenty airplanes have gone missing there as well. For some people, these disappearances defy explanation.

Add in tales of massive rogue waves and weird electromagnetic storms and claims of time warps, wormholes, gigantic methane bubbles and compasses that just won't work, and it seems the isosceles-shaped region whose vertexes are Bermuda, Miami and San

Juan, Puerto Rico, has earned its reputation for being a very mysterious and dangerous place.

The Bahamas, or at least most of them, are located inside the Bermuda Triangle; they provide just about all the landmass for what is generally considered a deep-sea phenomenon. Maybe it's no surprise, then, that the popular tropical islands are steeped in their own kind of strangeness.

Back in 1940, world-famous psychic Edgar Cayce predicted a piece of Atlantis would reveal itself in 1968 or 1969 off the east coast of North America. That prophecy apparently came true right on schedule when in 1968 dozens of enormous hand-cut stones were discovered in eighteen feet of water off the Bahamian island of Bimini. The stones, dubbed the Bimini Road, are thought to be at least four thousand years old.

There are more reports, from sources such as *Newsweek,* that a large pyramid may also have been found off Bimini and that other enigmatic underwater formations—some looking eerily like Stonehenge— have been discovered off Andros, the largest of the Bahamian islands. Some researchers believe all these artifacts are leftovers from the ancient civilization of Atlantis.

The Bahamas also hold legends of sea monsters that are part squid, part octopus and part shark. These monsters are said to live in the famous Bahamian blue holes, the unnatural-looking structures that dot the islands and whose origins are also in some dispute.

On those same islands, tales are plentiful of mysterious monkey-like creatures called Chickcharnies that are almost never seen but will attack and kill any human that approaches them.

The Bahamas are also rife with ghost stories, shamanism and the practice of voodoo. Ghost ships have been reported sailing the islands' waters for centuries.

Even Christopher Columbus wrote of seeing mysterious lights flashing on the horizon where the following day he would find land. That land, Samana Cay, is one of the Bahamas' outer islands.

Or how about the "electronic fog," that ethereal vapor that Bahamas-bound pilots have reported surrounding their aircraft and flying along with them in otherwise clear weather.

Then there's the most famous Bermuda Triangle case of all: the disappearance of Flight 19. On December 5, 1945, five Navy Avenger dive-bombers vanished while on a training mission. A large amphibious rescue plane sent to search for them also went missing. No wreckage was ever found of any of the airplanes, and no sign ever of the twenty-seven lost airmen. A lot of the drama that day involved the Bahamas.

A case can be made that, added all together, the strangest incidents inside the Bermuda Triangle—missing planes and ships, sea monsters, island monsters, baffling fog, ghosts and ghost ships, as well as the possible rising of Atlantis *and* many UFO sightings—occur in or near the Bahamas.

So it's odd, then, that in the middle of all this the U.S. Navy decided to build a *very* top-secret base.

The Official Word

This secret base is called the Atlantic Undersea Test and Evaluation Center, or AUTEC.

Built in 1954 on Andros Island, which is located just south of Bimini, according to the U.S. Navy's website, "AUTEC provides instrumented operational areas in a real world environment to satisfy research, development, test and evaluation requirements and operational performance assessment of war fighter readiness in support of the full spectrum of maritime warfare."

Classic military-speak . . .

But to understand the official explanation of AUTEC, one must first understand something called the Tongue of the Ocean. Commonly referred to as "Toto," it's a deep ocean basin located between the

Bahamian islands of Andros and New Providence. Twenty miles wide by 150 miles long, Toto is more than six thousand feet deep in some spots and has a relatively flat bottom.

The main support base for AUTEC is just astride Toto on the east coast of Andros Island, and the up-front, day-to-day activity there probably includes training new submarine commanders, testing new submarine-launched weapons systems and performing deepwater shakedown exercises for newly constructed submarines.

But in the Navy's own words, it has top-secret gear at AUTEC that can track up to sixty-three underwater targets simultaneously. Moreover, its significant land-based radar systems can spot anything flying within five hundred nautical miles of the facility, up to heights of seventy thousand feet.

How this technology aids in the "operational performance assessment of war fighter readiness in support of the full spectrum of maritime warfare" is a question best left to the experts. But simply having gear on hand that can find so many things under the water and so high in the sky has led some in the conspiracy-centric community to question the Navy's explanation about what really goes on at AUTEC.

Stealth Subs and Wormholes

Alternative theories abound. The Navy is secretly testing its new hypersonic rail gun at AUTEC, a fearsome electromagnetic weapon that can propel nonexplosive projectiles to speeds of up to six thousand miles per hour. The Navy has built massive underwater channels near AUTEC, tunnels that allow U.S. nuclear submarines to transit to an even more secret base inland on Andros Island. The Navy is working on making its nuclear submarines "invisible" through the latest in stealth technology.

Other theories are more way-out. AUTEC contains a star gate through which humans can travel to other star systems. Or has mini-wormholes that lead to another universe. Or Edgar Cayce was right and the Bahamas are actually the top of what was once Atlantis, and the Navy built AUTEC there to exploit the secrets of that ancient yet highly advanced civilization. That would make sense if you believed that Toto was created by some cataclysm worthy of Plato's Atlantis.

Mostly, though, AUTEC is thought to be the "Navy's Area 51."

UFOs over AUTEC

AUTEC shares some similarities with that top-secret Nevada base. Both are highly classified areas in very unusual locations. Practically nothing about what goes on at either place is available for public scrutiny. And both are sewn up extremely tight when it comes to security.

But also, much like Area 51, there's lot of UFO activity reported around AUTEC.

Here are just three of many puzzling cases:

THE LITTLE SUN

One highly unusual sighting, made in the 1980s, came from the teenage son of a Navy officer assigned to AUTEC. He and several friends were passing the time on a beach across from the top-secret facility. It was a very dark night, with no moon. Suddenly the beach, the water and a nearby harbor lit up as if it was daytime. Startled, the teens looked up to see a large circular object racing across the sky. They described it as moving very fast and being as bright as a little sun.

As they watched it go over their heads, the object suddenly broke up into four separate pieces. An instant later, these pieces completely disappeared, turning everything back to pitch-black again.

In all, the sighting lasted about thirty seconds.

SHRINKAGE REPORTED

As reported in August of 2010, this sighting was made by a civilian contractor who actually worked at AUTEC. He was walking the beach on his day off when he heard a helicopter approaching. Looking up at the aircraft as it went over, he noticed a strange object flying nearby only about five hundred feet off the water. The object was metallic in color and moving west to east. It had no lights and made no noise but was moving extremely fast. As the witness watched in amazement, the object came to a complete stop over a nearby island. It hung there for a moment, then suddenly shrank down to the size of a bird and vanished. As reported by the witness, all this happened on a perfectly clear day.

PARADE OF UFOs

In what might be the strangest UFO sighting over the Bahamas—and one of the strangest sightings anywhere—on the night of January 10, 1985, hundreds of people on New Providence Island, which is right across the Toto from AUTEC, observed what many of them described as a huge UFO leading an almost endless line of smaller, extremely bright objects across the sky.

This unexplainable array was estimated to be at least five thousand feet high. The lighted objects made no noise; in fact, many who saw them remarked on the eerie silence as the UFOs moved in perfect unison.

Writing the next day in the daily *Nassau Guardian*, columnist Rod Attrill said, "Whatever it was just kept coming and coming, with more and more bright lights following from behind. Within seconds a majestic procession was trailing across the sky, some brighter than the others, but all maintaining perfect speed and position in relation to the others."

The object leading this spectacle was described as bulbous and teardrop-shaped and to some witnesses nothing less than gigantic. One witness described it as being shiny metal with lights coming from the sides as if through windows. Another saw it as something akin to the body of a wingless plane with its cabin lights shining.

According to Attrill, based on the arc of sky covered by the objects, at five thousand feet high this lead object might have been a staggering half mile in length. Just how long the trail itself was proved to be incalculable, but it was hundreds of miles long at the very least.

No explanation has ever been given for the bizarre sighting.

From UFOs to USOs

But there's an unusual twist when it comes to AUTEC and UFOs, because in addition to many unexplained aerial sightings around the secret facility, there's been a lot of USO activity reported in the area as well.

USOs, as in unidentified submerged objects, are mysterious things seen in the ocean that, just like UFOs, can't be easily explained. Some researchers believe USOs are simply UFOs emerging from the ocean; others see them as a completely different phenomenon.

Physical descriptions of USOs are usually similar to those of UFOs—that is, disc-shaped or cylindrical objects, or glowing spheres. Frequently USOs are seen traveling at speeds faster than any Earth-based underwater vessels. Some have been reported moving as fast as five hundred knots when submerged—frankly impossible by our current laws of physics.

With thanks to www.abovetopsecret.com and the National UFO Reporting Center, below are some classic USO incidents that have happened near AUTEC over the years.

THE FREQUENT VISITOR

A particularly strange USO has been seen regularly off Andros Island in recent years. This craft resembles a UFO in shape and is said to move extremely fast, both under the water and in the air.

One witness saw this mystery craft close up while sailing off Andros. Spotting something moving along the surface of the water and thinking it was a whale, the witness steered toward it. But upon getting closer, he realized the object was hardly a whale. Rather, it was disc-shaped and made of gleaming metal in what he described as being an ultramodern design.

As the witness watched from just a short distance away, the craft rose to the surface of the water and began streaking along the surface at what the witness described as extremely high speed. It traveled like this for a few moments before lifting into the skies and disappearing overhead.

UNDER A STRANGE LAGOON

In an incident first reported in 2008, four people cruising on a yacht lost their radio and all electrical power soon after anchoring in a lagoon close to AUTEC. Moments later, the yacht's passengers saw five lights beneath the surface of the lagoon moving at incredible speed. The water was only about twenty feet deep and crystal clear. The witnesses estimated the lights were moving at least 150 knots or more, without leaving any discernible wake.

The lights themselves were bright yellow and a few feet in diameter. They made no noise but were performing incredible turns at full speed while submerged. Even though the yacht's captain had been involved in top-secret operations around the Caribbean while in the U.S. Navy, he later stated he'd never seen anything like these things.

At one point, the lights came together to form a solid disc-shaped metallic object. The astonished witnesses then saw this object rise out of the water and hover about twelve feet above the surface. It was so close to them that they could hear water dripping from it as it hung in the air. Then the object shot straight up and was gone in an instant.

As soon as the object disappeared, the yacht's electricity and radio came back on.

COLLISION COURSE

In the summer of 1984, the naval support ship USS *Yellowstone* was sailing past Andros Island, heading north toward its home port of Norfolk, Virginia. The night was calm. At three a.m. a sailor on watch noticed a green glowing spot on the water's surface about a mile in front of the ship. Brighter and more sharply defined than the bioluminescence most ships leave in their wakes, this spot was something out of the ordinary, the sailor realized.

By the time he radioed the ship's bridge with a warning, the green glow was less than a half mile away and coming on fast. The sailor thought it might be a research submarine, albeit a very speedy one. But whatever it was, he was convinced his ship was going to collide with it.

A few seconds later, the green flash went by the *Yellowstone*'s starboard side, not twenty feet away. The sailor got a clear look at it for the first time and said the glow was actually a sharply defined object about forty feet in diameter and perfectly circular in shape. It was at least ten feet thick and only about six feet below the water's surface. There was no sound and definitely no visible wake.

By the time others on the ship's bridge could react, the object was long gone.

The Navy and the Unknown

When people discuss UFO encounters with the U.S. military, usually the first and sometimes the only service mentioned is the U.S. Air Force.

This is understandable. The USAF is America's premier air service. Fighters, bombers, transport aircraft, ICBMs, satellites. The few official U.S. government investigations into UFOs were done by the Air Force. The famous, if ill-fated, Project Blue Book was an Air Force endeavor. Because UFOs are mostly an airborne mystery, they usually land in the lap of the U.S. Air Force.

Yet the U.S. Navy has had its own experiences with UFOs—extensive ones, both in the air and under the surface of the ocean.

During World War II, Navy pilots fighting the Japanese in the Pacific theater encountered foo fighters with almost the same frequency as their Army Air Force brethren did battling the Nazis over war-torn Europe. There are reports of UFOs trailing U.S. Navy ships during those Pacific operations and of more than a few USO incidents as well. (In fact, according to Keith Chester, author of *Strange Company*, the first in-depth accounting of aerial phenomenon observed during World War II, there may be many U.S. Navy Pacific theater reports regarding such things yet to be released or discovered in existing files.)

Navy pilots also encountered many UFOs during the Korean War (1950–1953), and Navy ships encountered them off the coast of that embattled nation as well. Yet it was the Air Force that took a lot of heat for not admitting U.S. military pilots were spotting UFOs during this so-called Forgotten War. In 1952, during a large NATO exercise called "Mainbrace," UFOs harassed a huge naval force that included a half dozen U.S. Navy aircraft carriers and many other surface ships. Hundreds if not thousands of sailors saw these objects—and many clear (but later impounded) photographs were taken of them. Yet officially the Navy has always kept quiet on the incident.

The same is true in one of the most spectacular sightings of the fifties, indeed in all of UFO history, when on February 10, 1956, a Navy transport plane carrying Navy aircrews home from Europe encountered an enormous flying disc a few hundred miles off the coast of Newfoundland. More than one hundred Navy personnel were on this plane and watched as the six-hundred-foot-long disc maneuvered effortlessly around their aircraft. Yet it was U.S. Air Force investigators who debriefed the passengers when the plane eventually landed in Gander.

Even more bizarre, one U.S. Navy aircraft carrier, the USS *Franklin D. Roosevelt,* was virtually haunted by UFOs throughout its entire service life, having many encounters with strange unidentified craft all over the world during its thirty years at sea. When the carrier was finally decommissioned, though, its logs were purged of any mention of these events.

History shows the U.S. Navy has simply managed to keep a low profile when it comes to unexplained aerial phenomena. Somehow, whatever America's oldest and largest military branch knows about UFOs it has been able to keep under tighter wraps than its cousin service, the U.S. Air Force.

What the Russians Say

The U.S. Navy's official position on the Bermuda Triangle, and by extension all the strange things that have happened in and around the Bahamas, will probably surprise no one.

This is from the Navy's own website (www.history .navy.mil): "The 'Bermuda Triangle' or 'Devil's Triangle' is an imaginary area located off the southeastern Atlantic coast of the United States, which is noted for a supposedly high incidence of unexplained disappearances of ships and aircraft. The U.S. Board of Geographic Names does not recognize the Bermuda Triangle as an official name. The U.S. Navy does not believe the Bermuda Triangle exists."

But the Russian Navy disagrees. According to the

Russian-based English-language news service www .rt.com, mysterious events happen in the Bermuda Triangle all the time. Rear Admiral Yury Beketov, a retired Russian Navy commander, tells of instruments showing unexplainable readings while his submarine was passing though the Bermuda Triangle and in waters near the Bahamas, disruptions he believed were caused by UFOs or USOs, or both.

"On several occasions the instruments gave readings of material objects moving at incredible speed below the water," Beketov said in an interview with rt.com. "Our calculations showed speeds of four hundred kilometers per hour [about 250 miles per hour]. Speeding so fast is a challenge even on the surface. But water resistance is much higher. There's only one explanation: The creatures who built these objects far surpass us in development."

Another Russian Navy veteran agreed. Naval intelligence officer Captain Igor Barklay said on rt.com, "Ocean UFOs often showed up wherever our or NATO fleets were concentrated. Near the Bahamas and Bermuda and Puerto Rico. They are most often seen in the deepest part of the Atlantic Ocean, in the southern part of the Bermuda Triangle and also in the Caribbean Sea."

Inside the Fence

One day, a handful of hunters on Andros Island mistakenly wandered into AUTEC's prohibited area. This was a big mistake. One moment they were alone, the next they were pounced on by armed men and their faces pushed into the ground. The armed men were Navy guards who had been under camouflage until the hunters practically walked on top of them. The hunters were handcuffed and brought to a nearby Navy facility where they were interrogated for hours by Navy officers. During this hair-raising time, the hunters were certain they were heading for jail.

The hunters insisted they had innocently walked into the prohibited zone, though, and eventually the Navy believed them.

But just change a few of the facts around and this could read like an encounter inside the perimeter of Area 51.

We spoke with an employee of a firm that works closely with U.S. intelligence agencies as well as the Department of Defense. He'd spent time at the AUTEC facility on Andros Island and painted a somewhat unexpected description of the place and its surroundings. The area around the base is actually a very run-down, poor region—more of a Third World situation and *not* at all the kind of tourist destination one associates with the Bahamas. He also spoke of many mangrove swamps and a little town close to the base that featured a few crumbling concrete houses with tin roofs and little else.

The AUTEC facility itself is huge, though—two square miles by one estimate—and obviously very well guarded. Inside is a different world. About a thousand people work there and the place features an officers' club, a snorkeling shop, sailboat rentals and a bowling alley. There's also a pub-type bar.

The only other distinguishing features he could discuss were a large radar set located on a beach nearby and a huge tree-trunk-sized cable that runs from the AUTEC facility across another beach and down into the Toto trench. And about a half mile from the base one of the Bahamas' mysterious blue holes can be found.

Noted UFO researcher and TV personality Bill Birnes has also been to AUTEC. He told us in an interview for this book, "The big thing isn't the base itself. It looks like any other base. The whole point of the base is the deep trench nearby. The base is a submarine base, and the trench is so deep, subs can get in and out without being detected. In other words they can submerge as deep as a sub can submerge, then get into the base without being tracked by satellites or anything else."

As for the many UFO and USO sightings that occur in the area, Birnes has an interesting hypothesis.

"Lots of people see lots of objects come out of the water and fly," he said. "My theory is that many of these are our own inventions. Because when you think about it, masking something to look like a UFO is the best way to mask a top-secret weapon."

In the End . . .

Where does all this leave us?

Put aside the deepwater naval training, the weapons testing, even the deep trench called Toto. The Navy has billions of dollars of sophisticated equipment at AUTEC, equipment that by their own admission can track multiple targets under the water and very high in the sky.

At the same time, people see lots of UFOs and USOs in the vicinity of AUTEC and have been seeing them there for quite some time. So, could the Navy really be unaware of all this strange activity going on right in its own watery backyard? With so many ways to find things underwater or flying in the sky, many UFO researchers think that would be impossible.

This leads to theories, then, maybe not so crazy, that the Navy is very much aware of what's happening in these mysterious waters, and that just like their brethren at Area 51, they might even be the cause of all the strangeness.

Maybe Bill Birnes's theory is right. Maybe the Navy has secret weapons at AUTEC that are disguised to look like UFOs and USOs. That would answer the question of why so many unusual things are seen around Andros Island and Toto. It would also provide a perfect cover story—because after all the fringe press the Bermuda Triangle gets, who is going to be surprised when someone sees yet another UFO or USO inside it?

But this also begs another question: What kind of secret weapons could these things be if the Navy has the ability to disguise them to look like UFOs and USOs, objects that are seen doing some pretty fantastic things?

Until we find out for sure, the biggest mystery of the Bermuda Triangle might still be why the Navy chose to build their secret base there in the first place.

9

The Mystery of Ong's Hat

Haunted Woods

The Pine Barrens is a peculiar place in a peculiar location.

A large forest that takes up a significant portion of New Jersey's southern coast, the Pine Barrens stretches from Lakehurst in the north to Cape May in the south. The Garden State Parkway and Atlantic City Expressway run right through it and it's close to both New York City and Philadelphia. Yet many people are unaware the Pine Barrens exists. Fifteen hundred square miles of woods that are undeveloped, rural and almost devoid of people, in the midst of one of the most heavily traveled, densely populated parts of the United States.

The chief reason no one lives there is the soil. The locals call it "sugar sand." Highly acidic, it holds almost none of the essential nutrients needed for growing agricultural products. When European settlers first arrived in the area in the seventeenth century, they found their crops would not grow in the Barrens' soil, so the vast majority of them went elsewhere.

That doesn't mean *nothing* grows there, though. Just the opposite. The Pine Barrens boasts some very strange flora: four-foot-tall "pygmy" pines, rare orchids, *ultra*-rare curly grass ferns . . . and carnivorous plants. Lots of them.

There are also rattlesnakes here.

And ghosts.

And at least one monster.

The New Jersey Devil (Among Others)

It all started with a woman named Mrs. Leeds. She lived deep inside the Pine Barrens and around 1735 gave birth to an infant who was right out of a horror movie.

The baby boy was her thirteenth child, and cursed

from the beginning. One account has the infant attacking its mother and the attending nurses at the moment of birth, before escaping up a chimney. Variously described as having the body of a serpent, the head of a horse, the wings of a bat and a devil's forked tail, that monstrous toddler became the New Jersey Devil and has been haunting the Pine Barrens ever since.

But when it comes to the paranormal, the Barrens can be a pretty crowded place.

According to legend, close-by Barnegat Bay is one of the resting places of the notorious pirate Captain Kidd. Locals tell tales of Kidd's ghost endlessly walking the beach, looking for his buried treasure, often minus his head.

There is the Barrens' Black Dog, the ethereal canine that roams the forests from Absecon Island to Barnegat Bay. The story goes that pirates hiding on Absecon Island murdered the crew of a merchant ship, including the ship's cabin boy and his dog. The dog now walks the Barrens searching for his long-lost owner.

Another Pine Barrens ghost eternally pining for a loved one is the Golden Haired Girl. She is frequently seen dressed all in white and looking forlornly out to sea for a drowned lover who will never return.

Then there is the White Stag. This ghostly deer is a helpful spirit known to lead travelers lost in the Pine

Barrens to safety. The stag also has the reputation for preventing mishaps, on one occasion stopping a stage-coach from falling into a raging river by blocking its path to a bridge that had been washed away.

According to legend, if you see the White Stag, it's supposed to be good luck, not a bad thing with Atlantic City just a few miles down the road.

No surprise that the Pine Barrens also has an authentic ghost town.

It's called Ong's Hat. It began as a small village sometime in the nineteenth century and was never more than a handful of buildings at the intersection of two roads. Its odd name came from an incident in which a local man named Jacob Ong threw his hat up in the air, for reasons unknown, only to get it stuck in a tree. The hat remained entangled in the branches for so long, whenever anyone passed through the tiny village and saw Ong's Hat, they knew where they were.

By the 1920s, just about all traces of Ong's Hat village had disappeared. The tiny community continued to appear on maps though little more than just a name in the middle of this odd, out-of-place wilderness.

Yet, it was here, in Ong's Hat, that a small group of

scientists did nothing less than open a door to another dimension.

But it was also here, soon after these scientists made this startling breakthrough, that the U.S. military raided Ong's Hat, burned down the group's laboratory, and by most accounts killed all of its members.

The Universe Next Door . . .

The Pine Barrens has been described as a perfect place for a UFO landing.

Perhaps that's what first attracted twin brother and sister Frank and Althea Dobbs to it. The Dobbs twins spent their early childhood on a UFO cult commune in rural Texas, a collective founded by their father.

It's said that, as undergraduates at the University of Texas, the twins came up with a series of equations they were certain contained the seeds of a new science, which they called Cognitive Chaos. In this, they theorized that people could not only heal themselves of any affliction and put an end to aging, but they could also travel to other dimensions with very little physical effort.

Though later enrolling in Princeton University, the

twins soon became disenchanted with that institution's strict academic policies. Nevertheless, they worked diligently to prove interdimensional travel was attainable through the science of Cognitive Chaos. But after submitting a thesis on the subject sometime in the mid-1980s, the twins were immediately expelled from the famous Ivy League school.

Determined to pursue their studies, the twins bought a rundown Airstream trailer and drove it to Ong's Hat. Hidden away in the Barrens, they constructed a crude laboratory inside the trailer and went back to work on their theory. Sources of income presented themselves via the sale of certain illegal agricultural items (translation: pot), and soon a commune of sorts rose up around the tiny settlement. Many members of this collective were local runaways.

Time passed. The twins did their science and the commune grew. The tiny community eventually became known as the Institute of Chaos Studies, or ICS. Two more scientists joined the twins in their studies. Contact with other underground experts in various related fields was made as well.

As it turned out, even the twins were shocked with the advances they were making. Within a year just about everything their equations had predicted had come true.

A little more than three years later, now in the late 1980s, they made their breakthrough: They created a device they called the "Gate."

On one side was the old trailer. On the other side, another dimension.

The group built a capsule, a vehicle to travel through the Gate. One of their commune members, a person described as a juvenile delinquent, volunteered to be the first one to go through the portal.

On the big day, and apparently at the right moment, this capsule vanished from the laboratory, with the young volunteer strapped inside.

The group began to panic, not quite sure what they had done. But seven minutes after it disappeared, the capsule reappeared. Its passenger was still inside, alive, unscathed and beside himself with delight.

The passenger reported that he had indeed passed through to another dimension. In fact, on the other side of the Gate he found . . . the Pine Barrens. It was not the one he'd traveled from, but a place very much the same, except for one thing: In this dimension, there were no humans anywhere. Same place, just no people.

This was just the beginning for the visionaries at Ong's Hat. As time went on, the group perfected ways of visiting more dimensions and they journeyed to these places en masse. But they always kept a soft spot for the first place their volunteer had traveled to.

They set up a sort of alternate colony in this different Pine Barrens and began spending most of their time there. During this period, they would often come back to our present dimension to get necessities like computer parts and books, coffee and beer. There was never any intention to "escape" to the other dimension and totally leave this one behind. They were travelers; they liked going back and forth.

Then something went wrong.

Details are sketchy, but either word got out to the powers that be of what the group was doing or the government suspected the group had something to do with a dangerous chemical leak at nearby Fort Dix. In any case, the commune at Ong's Hat was stormed by members of a Delta Force team operating out of Fort Dix, with the New Jersey State Police SWAT team backing them up.

The Delta Force troops arrived over Ong's Hat in helicopters and rappelled down long ropes, carrying automatic weapons. In an action eerily similar to the Waco Branch Davidian massacre, the Ong's Hat commune was set ablaze and burned to the ground. As many as seven residents were either killed or disappeared in the raid and all evidence of the settlement was bulldozed over, finally eliminating Ong's Hat for good.

Yet none of this was ever reported in the media.

Because maybe none of it ever happened.

The Legend of a Legend

Is the story of Ong's Hat real?

That depends on how you define *real*.

But ruse or not, the genesis of how the tale came about is almost as strange as the tale itself.

It's widely believed that the story of the ICS commune at Ong's Hat was created in the early 1990s by author Joseph Matheny in an attempt to insert it into the collective consciousness of a then budding World Wide Web. To this end, Matheny and others planted different pieces of the saga disguised as true items on bulletin boards and in early Internet zines.

It took a while, but eventually others began responding to these postings, adding their own artifacts, from personal anecdotes to audio and even video recordings. As more people added pieces, Matheny's original narrative began to build like an incredibly involved kids' game of telephone.

In other words, Ong's Hat was a legend that grew

not by numerous retellings around a campfire but by numerous retellings on the Internet. In fact, it was probably the Net's first real myth.

But . . . because the story of Ong's Hat induced such extraordinary responses from so many participants, many of them believed the story was real. This phenomenon is based on the theory that the reason strange things "exist" in this world—ghosts, monsters, aliens— is because so many people think about them, and with so many people thinking about them then they must be real.

Michael Kinsella, author of the book *Legend-Tripping Online: Supernatural Folklore and the Search for Ong's Hat* (University Press of Mississippi, 2011), told this writer, "Modern supernatural legends thrive because they not only describe tantalizing other worlds and possibilities, but they also serve as templates for all kinds of ostensive acts that may lead to all kinds of ostensibly supernatural experiences."

In fact, even though some reports say Matheny called the story to a halt in 2001, at one time, hard-core participants actually embarked on pilgrimages to the Pine Barrens, hoping to find (and maybe finding) the interdimensional portal at Ong's Hat.

But was it *really* "real"? The twins, the trailer, the capsule, the "other Pine Barrens"? Did they really "exist"?

We'll leave that answer to the man who created the greatest detective of all time.

There is a story about a question once asked of Sir Arthur Conan Doyle, author of the Sherlock Holmes mystery novels.

The question was, "Is Sherlock Holmes a real person—or is he fiction?"

Without missing a beat, Doyle replied, "Yes."

PART TWO

Other Secrets Around the World

10

Looking for Britain's Area 51

This Strange Island

At a British air force base called RAF Montrose, there's an old radio from the 1940s that still plays speeches by Winston Churchill and the music of Glenn Miller, broadcasts from more than seventy years ago.

Dozens of people have witnessed this over the years. The broadcasts come on at random times and last for thirty minutes or more. Technicians who disassembled the radio and found nothing unusual about its construction are baffled at how it can still pick up live shows from the dark days of World War II. Even more remarkable, the old radio does this without being plugged into any power source.

But haunted radios are just the beginning when discussing what a strange place Great Britain is.

The British Isles are studded with puzzling circles of standing stones. Stonehenge is the most well known of these structures, but there are hundreds of others just as mystifying scattered throughout the UK. Britain also boasts more than *ten thousand* round barrows, which are another kind of ancient "rock art." No one knows for certain who built them or why, only that most of them date back more than five thousand years.

Great Britain is also crisscrossed with "ley lines," alignments that some believe run straight through many of the ancient monuments, megaliths and other mysterious sites. In other words, all of those henges and round barrows, as well as places disguised to look like natural ridgelines or river crossings, are not there randomly; rather, they were purposely lined up and connected by ley lines, as if built under one vast ancient design. What's more, many believe that ley lines themselves are magical and contain special psychic energy.

Then there are the crop circles. Between 1970 and 2000, two dozen countries reported crop circles suddenly appearing in their midst—ten thousand circles in all. Of these, *90 percent* were found just in southern

So many cars have been chased by flying objects along Australia's Nullarbor Plain, one local government erected a sign warning motorists to "Beware of the UFOs." *Photo by Robert Paul Van Beets*

Some researchers say more UFOs are reported over the Falkirk section of Scotland than anywhere else in the world. But this region is also home to the famous Rosslyn Chapel of Knights Templar lore. Could there be a connection? *Photo by S. Duffett*

For nearly ten years, the United States Air Force kept the F-117 Stealth Fighter hidden from friend and foe alike by flying it out of Nevada's highly classified Tonopah Test Range, a place even more secret than its desert neighbor, Area 51.

Photo by Trekandshoot

Explorers entering Russia's M-Triangle in the Ural Mountains report hundreds of strange incidents, including multiple UFO sightings; meetings with aliens; unexplained noises; bizarre star formations and, in some cases, a fantastic increase in intellect, leading some to claim the area is a piece of "Heaven on Earth."

Photo by Alx Yago

In 1962, China and India fought a war over the Aksai Chin region in the Himalayan Mountains. Some UFO researchers are convinced both sides suspected a UFO base was located somewhere close to the battleground. *Photo by Daniel Prudek*

Though searching for it is like chasing Alice through Wonderland, the well-hidden British Area 51 might be located surprisingly close to the equally mysterious ancient formation at Stonehenge. *Photo by David Maska*

Richard Nixon and Jackie Gleason were unlikely friends who shared a love of golf. When Nixon learned of Gleason's long fascination with UFOs, the president is said to have brought the comedian to Homestead Air Force Base in Florida to show him the wreckage of a flying disc, as well as the bodies of its occupants.

Photo © by Everett Collection Historical / Alamy

Many of America's nuclear submarines undergo testing at AUTEC, the mysterious Caribbean facility that's been called the United States Navy's Area 51.

Photo courtesy of the United States Navy

England alone, and many of these materialized near ancient monuments such as Stonehenge.

Other weirdness in Britain includes: Huge black cats frequently spotted roaming the countryside. Large black dogs, also known as Hell Hounds, that sometimes take on the shape of humans. Creatures called Mud Maids that lurk in the reeds along riverbanks, trying to lure victims to join them. And the Fear Liath Mòr ("big gray man"), a huge, hairy humanoid that is basically Britain's Bigfoot.

There are ghosts aplenty, too. One city, Chester, located in northwest England, has been seriously haunted ever since it was built in AD 70—that's nearly two thousand years of ghosts, witches, devils and imps running amok. Nearby is Armboth Fell, the location of Thirlmere Lake, where the echo of ghostly bells is frequently heard coming from beneath the water. At Chillingham Castle in Northumberland the cries of a young boy buried alive in one of its walls can still be heard at night. And in Sudbury where once stood Borley Rectory, the site of a fatal love affair between a monk and a nun, ghost sightings are still so frequent that the grounds are considered the most haunted place in England.

Then, of course, there's the Loch Ness Monster.

UFOs Over the UK

It's no surprise, then, that UFOs are fairly common in the United Kingdom as well. Tens of thousands of reports of unidentified flying objects have been made over the years, surprisingly large numbers for a nation that's relatively small.

But like staging a Shakespeare play or interpreting a poem by Blake, UFO episodes in the UK are rarely simple affairs. The string of strange events that took place around Dyfed, Wales, in 1977 is a case in point. It had a little bit of everything: multiple UFO sightings, glowing balls of light chasing cars and alien beings walking around the countryside—and even some peeping in people's windows. Other weird goings-on included: one particular UFO that liked to hover over a schoolhouse; TVs, radios and cars that stopped working for no reason; and the teleportation of a large number of cattle from one place to another.

The village of Warminster, near Stonehenge, provides another example. In the 1960s after strange, unexplained sounds began to be heard near the town, UFOs arrived in the area—and never left. By one estimate, more than *five thousand* unidentified flying objects have been spotted near Warminster over the ensuing decades.

There's also an interesting place in Scotland called Bonnybridge—but more on that later.

But Where to Look?

These days, Great Britain's military is world class, perhaps second in some respects only to the United States. Like the States, the UK is regularly designing and developing new, highly classified weapons of its own. And it stands to reason that in doing so, the British would need a place to test and in many cases fly these new weapons, while at the same time keeping them absolutely top secret.

And, because these things seem to go hand in hand, if, as some UFO researchers claim, the British have crashed UFOs in their possession, or pieces of them, they would need a supersecret place to hide and examine these things as well.

In other words, they would need a place like America's Area 51.

But do the British even have one?

Such a place is rarely, if ever, mentioned in the British media. There are no websites devoted to a British version of Groom Lake. We could find no books that expound on the topic.

So, it's a bit of a challenge. Groom Lake, Tonopah, Homestead, AUTEC—we know *where* these places are. What we don't know is what exactly is done behind their heavily guarded gates or what if any is their true connection to UFOs But in Britain, the challenge is finding the "candidate" top-secret base in the first place. That one signature secret location that is entwined with the UFO mystery.

So, we put the question to Nick Pope, the well-known author, journalist and TV personality who, among other things, once investigated UFOs for Britain's Ministry of Defence.

Because he is still bound by the Official Secrets Act, Pope had to be careful not to divulge any classified or sensitive information. But he did educate us a bit via an interview for this book.

"As to whether the UK has its very own 'Area 51,'" he said, "the answer is—as you'd expect from a former civil servant—yes and no. Clearly there are different ranges and danger areas where training is carried out and where various aircraft, UAVs and weapons systems are trialed. But obviously there's nothing on the scale of the real Area 51. Flying at supersonic speed, an aircraft can cross the UK from west coast to east in a few minutes. Contrast that with the size of the continental U.S., where vast areas are uninhabited or very sparsely populated. Simply put, there's not much room in the UK for a British Area 51. That said, though, rumors persist about a couple of sites."

Actually more than a couple. Ask around further and you'll get a number of different candidates and, no surprise, all of them have typical British names: High Wycombe, Fylingdales, Aldermaston, Boscombe Down, Rudloe Manor, Menwith Hill, Hack Green, Warton Aerodrome, Porton Down. There's even a phantom English town named Argleton that seems highly suspicious. And interestingly enough, almost all of these places have some kind of UFO connection.

But which place is the most likely candidate for a British Groom Lake?

Though searching them all would be an exercise not unlike chasing Alice through Wonderland, here are the top ten places Britain's Area 51 might be found:

10. RAF HIGH WYCOMBE

Located near the village of Walters Ash in Buckinghamshire, RAF High Wycombe provides support for the Royal Air Force's Air Command. However, in 2009 the British Ministry of Defence transferred all of its civilian UFO researchers here from their former offices in London. RAF High Wycombe is where all UFO sightings in the UK are now "officially" investigated.

At the time, Nick Pope hailed this change of location, optimistic it would mean not only more investigations but more thoroughness in those investigations. He especially hoped the RAF would pay particular attention to sightings from pilots and cases where UFOs are tracked on radar.

But a spokesman for the Ministry of Defence spoiled all the fun by telling the media, "The MOD examines UFO reports solely to establish whether UK airspace may have been compromised by hostile or unauthorized military activity, [and] unless there is evidence of a potential threat, there is no attempt to identify the nature of each sighting reported."

Still, hats off to the British for at least acknowledging they have an official body looking into UFO sightings.

But apparently that's all they do regarding UFOs at RAF High Wycombe. It's unlikely this is Britain's Area 51.

9. RAF FYLINGDALES

At least on the face of it, Fylingdales is a simple radar base located in Yorkshire. But though maintained by the RAF, the installation is actually part of the United States' Ballistic Missile Early Warning System, one of three secret facilities the United States and the UK share. (The other two are in Greenland and Alaska.)

Nick Pope told us, "RAF Fylingdales is an early warning station and has a secondary space surveillance role, detecting, tracking and reporting satellite launches and orbits. Its powerful space-tracking radar systems also monitor space debris and the reentry into the Earth's atmosphere of rockets. Because satellites and rocket reentries are sometimes mistaken for UFOs and because the space-tracking radar system would detect anything unusual at great distance, the base had a role in MOD's UFO project."

Fylingdales is located relatively close to a place called RAF Menwith Hill, and the two bases have similar functions. Both have also been the sites of protests by people who believe the facilities encourage the militarization of outer space. Perhaps this is another reason why Fylingdales is frequently suspected of being more than just a radar station on steroids and somehow has links to UFOs.

However, there has never been a huge amount of UFO activity reported in the Fylingdales area over the years, nor have there been any rumors of crashed UFOs hidden within. As these things seem necessary to be crowned UK's Area 51, RAF Fylingdales is probably not the place we're looking for.

8. RAF MENWITH HILL

The Menwith Hill RAF installation, also located in Yorkshire, has an even closer working relationship with the United States. Simply put, Menwith Hill has been called the world's largest spy base. It is also considered the most secretive facility in Britain. Actually run by the NSA, America's principal intelligence agency, the base is so secret that few people knew it even existed until a few years ago.

Menwith Hill is surrounded by a high fence, barbed wire, dozens of cameras and a small army of security guards. And apparently the number of igloo-shaped radomes on the site keeps increasing, meaning the base, and whatever it is doing, keeps getting bigger and bigger.

Rumors say these newer radomes are linked directly to America's spy satellites, orbiting spacecraft that can also pick up UFOs.

But Nick Pope said of the base, "To the best of my knowledge, RAF Menwith Hill has no role in UFO research or investigation. I only visited the base once in my twenty-one-year MOD career, not during my time on the UFO project, but during a subsequent posting to the Directorate of Defense Security. Despite the extreme secrecy and sensitivities, I wouldn't describe RAF Menwith Hill as 'Britain's Area 51' any more than I would label RAF Fylingdales in this way."

7. ALDERMASTON

Aldermaston is the only place in Great Britain where nuclear weapons are manufactured. Located in Berkshire, in the southern part of England, it's the headquarters of Britain's so-called Atomic Weapons Establishment, known by the great acronym AWE.

Aldermaston is responsible for the design and manufacture of the Trident submarine intercontinental ballistic missile system, as well as the dismantling and decommissioning of Britain's obsolete nuclear weapons.

(Scientists at Aldermaston also have an unusual side job: They monitor earth movements all over the world to detect any secret underground nuclear explosions. This is part of the effort to regulate nuclear testing as spelled out in the 1996 Comprehensive Nuclear-Test-Ban Treaty.)

It's been well established by serious UFO researchers that UFO activity picks up around facilities built for the purposes of using nuclear power, be they electrical generating stations, weapons factories or weapons sites themselves. And there is a strange UFO incident linked to Aldermaston.

According to a file released by Britain's National Archives, one morning in 1967, residents of southern England awoke to find "six small beeping UFOs lying in a perfect line from the Isle of Sheppey to the Bristol Channel."

While a bomb disposal unit blew up one of these UFOs, another was airlifted to Aldermaston, where both army and MOD's intelligence units were being mobilized for what was considered at the time to be a real alien invasion.

It was later discovered, though, that engineering students at Farnborough College of Technology had constructed and positioned the UFOs in a bid to raise money for charity.

Other than that fire drill, Aldermaston seems to be relatively UFO-free, inside and out.

6. RAF HACK GREEN

RAF Hack Green has a sterling record for defending Great Britain during World War II. Equipped with radar, searchlights and fighter plane control, the Hack Green base was assigned to defend all the airspace between Birmingham and Liverpool from German attack before, during and after the Blitz.

Later on, during the Cold War, a top-secret defense weapon known as the WE-I7 was placed at Hack Green. Used to detect a nuclear missile heading toward the UK, it would automatically fire a retaliation missile in such an attack, a missile that itself was nuclear armed. Luckily, the WE-I7 missile never had to be used.

Sometime in the 1960s, a bunker at Hack Green was rebuilt and designated as a Regional Government Headquarters, one of many similar sites around the UK where government officials could seek shelter in the event of a nuclear attack and continue operating in its aftermath.

Numerous UFO sightings were reported in the vicinity of Hack Green during the 1980s. One incident involved something landing in a nearby pasture and killing three cows, immediately attracting the interest of Britain's intelligence agencies. And though the base officially closed in 1992, there were whispers about a vast underground bunker still operating on the site, a place connected to a conspiracy involving the British military and UFOs.

This mysteriously inaccessible area at Hack Green was called "Underground Level 1," and it contributed greatly to talk of an ET connection. However, a para-science research group investigated the site and found that any closed-off areas were actually being used for such mundane things as storage bins. Nor were any vast tunnels or chambers found as the conspiracy theorists had speculated.

This led the research group to conclude that the British government was no longer involved there.

So, cross Hack Green off the list.

5. RAF BOSCOMBE DOWN

RAF Boscombe Down is located in Wiltshire, not far from Stonehenge.

It is an interesting entry on the countdown list for several reasons. Just as Area 51 boasts one of the longest military runways in the United States, Boscombe Down maintains the longest military runway in the UK.

And just like Groom Lake, Boscombe Down's primary role is to support flight trials of new aircraft, airborne equipment and weapons. Plus it's home to a fleet of classified test aircraft, just as Area 51 is known to be.

Best of all, there was a strange incident at the base in September 1994 that links it directly to Area 51.

One night a mysterious aircraft was taking off from Boscombe Down when something went wrong. Either the aircraft crashed shortly after takeoff or it never made it off the ground at all. Witnesses later saw the wreckage at the end of the runway, covered in tarpaulins and surrounded by security personnel. For the next twenty-four hours activity at the base was intense. This included not only the arrival of a huge U.S. Air Force C-5 cargo plane but also the sighting of an airplane extremely similar to a "Janet Airlines" aircraft, which was seen landing at the base. Janet Airlines is the nickname for the air service that transports employees of Area 51 from Las Vegas to Groom Lake. It is widely believed to be owned and operated by the CIA. If a plane like that landed at Boscombe Down, something very odd must have been going on.

Adding to the mystery, on February 7, 1995, there was a report of a UFO surrounded by flashing lights landing in a field close to Boscombe Down.

But unfortunately that's the extent of UFO activity reported in or around the base.

Still, it's an interesting candidate.

4. ENGLAND'S PHANTOM TOWN

Until fairly recently, the village of Argleton, located in West Lancashire, England, had everything expected of a small British town.

Located off the A59, it had its own postal code— L39. The town was right there on Google Maps. It had want ads for new jobs, an extensive real estate listing, its own weather report and even a local dating service.

Trouble was, Argleton didn't exist. In fact, it never existed. Anyone who went to the land coordinates given on Google would have found an empty field.

How did this happen? *Why* did it happen?

One British newspaper said, "Google and the company that supplies its mapping data are unable to explain the presence of the phantom town."

Then—just as mysteriously as it materialized, the town disappeared.

As of January 30, 2010, Argleton had been wiped from all Google maps.

What does all this mean? There's no way of telling. There were no UFOs reported near the place, nor were there any signs of military activity in the area.

However, as our Spook friend told us, "What better place for a government to hide secrets than in a town that never existed?"

3. WARTON AERODROME

Of the place called Warton Aerodrome, Nick Pope said, "It's a real potential candidate for Britain's Area 51."

Located in the northwest part of England a few miles north of Preston, Warton Aerodrome (as it's listed on maps) in reality is the place where global defense giant BAE Systems is rumored to be working on the next generation of top-secret stealth aircraft, stealth drones and God knows what else.

Nick Pope told us there have been numerous UFO sightings in the area, leading some people to suggest that the new aircraft being developed at Warton Aerodrome have been back-engineered from crashed alien spacecraft.

In fact, BAE found itself uncomfortably in the spotlight in 2009 when a "UFO" was reported to have hit a wind turbine in Lincolnshire. Ministry of Defence insiders suggested that a new stealth drone, code-named Taranis, might have been involved. This could mean that what people are seeing flying around Warton Aerodrome might be strange but are also of very earthly manufacture.

Still, Warton Aerodrome makes the top three.

2. PORTON DOWN

Ostensibly a government and military science park, Porton Down is located near Salisbury in Wiltshire, again in the same neighborhood as Stonehenge.

Most maps of Porton Down show "Danger Area" signs surrounding the entire complex, which at seven thousand acres is huge by British standards, and the place is definitely one of Britain's most sensitive military facilities. Highly classified research, including defense against chemical, biological, radiological and nuclear warfare is conducted there—and that's just the work we know about.

More important, though, Porton Down is associated with one of the strangest UFO stories in British history.

Well told by Nick Redfern in *Keep Out,* this tale began on the night of January 23, 1974, when rumors spread through the British military establishment that something very strange had happened near a mountain in North Wales.

That night, one particular British military unit in the south of England was put on alert and told to await orders. Soon enough, this unit was instructed to head toward Birmingham, England's second-largest city, under strict instructions to make the journey as quietly as possible.

As the unit was approaching Birmingham, it received a second set of orders—it was now directed to head at full speed toward North Wales. This they did—only to be stopped again on the outskirts of (of all places) the seriously haunted city of Chester. At this point they were told they would soon be participating in a military exercise and to stand by. But then the orders were changed yet again and they were told to drive to a place in northeastern Wales called Llangollen.

There was much going on in this tiny village when the soldiers arrived. Townspeople rushing about, lots of British officers and officials in evidence, all while British fighter jets constantly flew overhead. Shortly before midnight, the soldiers were told to load two large boxes onto their vehicles. Then they were told to bring these boxes to Porton Down.

Arriving at the highly secret facility sometime later, the soldiers delivered the boxes, still sealed, as ordered. But for some reason, the people at Porton Down opened the boxes in front of the soldiers, to reveal that they were actually coffins. Inside were two beings not of this Earth.

The source of the story described the bodies as being five to six feet in height and humanoid in shape, but almost skeletal in appearance. The source said he and his other soldiers put two and two together and realized they'd just delivered the bodies of two ETs who'd been killed in a UFO crash up in Wales. This theory was later confirmed by other members of their unit. The source was also told that not all of the saucer's occupants had died in the crash.

But the story gets better. As noted by Nick Redfern, later on the British Ministry of Defence, in a bizarrely uncharacteristic move, agreed to provide the BBC with an enormous amount of technical assistance in the making of a science-fiction series called *Invasion: Earth*.

As its title indicates, the series was about aliens attacking our planet, which made the MOD's support so unusual, because many times up to that point the MOD had gone out of its way to discourage the idea that UFOs existed. Now, suddenly, it was throwing its considerable weight behind a series depicting UFOs invading Earth. Redfern said this got many people thinking. Was the British government attempting to get the general public in line with the idea of fighting against malevolent ETs?

But maybe the strangest thing of all is that in *Invasion: Earth* a number of aliens retrieved from a crashed UFO are taken to . . . yes, Porton Down.

"Did the MOD know something we didn't?" Redfern remembered wondering at the time.

Either way this is strange, top to bottom.

But does this mean that Porton Down is Britain's Area 51? It seems to fit the two main requirements: It's a highly classified facility, and it has a connection to UFOs—in this case, a highly dramatic one.

But there is one more place to look. . . .

1. RAF RUDLOE MANOR
Rudloe Manor is an innocuous-looking RAF facility located in Wiltshire, again not far from Stonehenge.

There are no three-mile-long runways or rows of modern fighter planes here. Rudloe Manor looks just as its name implies: It is a stately British estate that conjures up visions of fox hunts, the gentry, crumpets and tea.

But like Megiddo, Rudloe Manor is good at hiding its secrets. Below it is nothing less than a vast underground city made up of hundreds of man-made caverns and tunnels, with room enough for an aircraft factory and a train station that secretly connects to the Tube in London, eighty-five miles away.

But most interesting, Rudloe Manor has an intriguing, if unusual, connection to UFOs that puts it at the top of this list.

We know there's a city beneath Rudloe Manor because a British TV news crew was allowed to film it back in 2000.

This subterranean city's origins date to World War II. During the early days of the conflict, the British built a huge network of bunkers under Rudloe Manor and the surrounding countryside. After coring out more than two million square feet of space, the British first built and then stored military hardware inside, far away from falling German bombs. The underground facility was so vast that an entire working aircraft factory was located there.

When Britain's Sky News was allowed into the

underground city, sixty years after it was built, its film crew reported on the caverns, the miles of tunnels that included underground railway stations and a decommissioned nuclear command bunker. Sky News also transmitted pictures of massive underground ventilation fans the size of modern jet engines.

While the British military insists that much of Rudloe Manor, or at least what's below it, is no longer active, rumors over the years suggest the base is actually a key part of Britain's secret investigation into UFOs.

Yet, Nick Pope told us, "Rudloe Manor has been dubbed Britain's Area 51, but I think that's more because the old RAF base there was involved in MOD's UFO project. There were rumors of secret, underground bases there. There certainly were a whole series of underground tunnels, bunkers and other facilities in the area, but this dates back to the Second World War and the Cold War and has nothing to do with UFOs or aliens."

Of course we must remember that Nick is still under the Official Secrets Act. If Rudloe Manor were still active, would he be able to tell us?

Which brings us to what the late editor of *UFO Magazine*, Graham Birdsall, said about this subject of supposedly inactive British bases: "As recently as 1997, I learned how communication technicians were operating from supposedly mothballed bases in the UK whose location and means of access to underground facilities below must remain confidential. However, I can categorically state that one such access point was described as being in the middle of a plain-looking field."

This is where our search begins to make a little more sense. If the British military truly has an Area 51, as we think of it, clearly it has no wide and empty desert to put it in. A mountain installation is a possibility, except that the UK's tallest mountains are in Scotland, where access is a little difficult and where it's cold and damp a lot of the time. In fact, operating out of a mountain anywhere on the British Isles would seem to be a cold and damp proposition.

But if there was ever a place where the British, on their strange little island, could do all kinds of secret things—including work on crashed UFOs or studies relating to the ET puzzle—totally out of the public and prying eyes, what better place would that be than underground? And what better place to do it underground than in an underground city that's already built?

There are other clues out there that, when put together, seem to point us in a certain direction.

What follows are five reasons we think our search for Britain's Area 51 ends at Rudloe Manor:

1. THE MINISTER WHO ASKED

On October 17, 1996, Martin Redmond, a member of the British government, asked a number of questions in Parliament about the UK's investigations into UFOs.

Redmond was subsequently given information on some classified military units, including one known as the Provost and Security Services, or P&SS. While on the face of it P&SS is an elite RAF unit that deals with things such as vetting of employees and issuing passes and so on, as Nick Redfern points out, some P&SS investigators are also trained counterintelligence agents who investigate such hot items as theft of classified material, acts of espionage and countless other things not known to the British public. Think of them as the RAF's version of the FBI.

And as Parliament Member Redmond found out during his UFO inquiry, at the time, P&SS had its headquarters at . . . Rudloe Manor.

2. THE "GOOD" REPORT

As detailed in *Keep Out*, several years earlier, in 1991, UFO researcher Timothy Good wrote of talking to a former P&SS special investigator who had specific knowledge of P&SS involvement with UFOs. Good's story was backed up by a second former P&SS counterintelligence agent who stated that when he worked for

P&SS, he had access to just about all of the agency's top-secret files—except those he understood dealt with UFOs. He said an armed guard stood watch over the door where those files were kept . . . at Rudloe Manor.

3. THE CIVILIAN WHO CALLED WHITEHALL

In 1984, a British citizen who'd witnessed a UFO sighting phoned Whitehall, the center of the British government, to report the event. This was a bold move, the equivalent of someone in the United States calling the White House.

But instead of giving the caller the number to the MOD Main Building—where we know civilian investigators were keeping track of UFO sightings at the time—the caller was given, possibly by mistake, the number for . . . Rudloe Manor.

4. HOW LOW IS "LOW"?

One unusual aspect of P&SS duties is its overseeing of the British government's "Low-Flying Aircraft Complaints" department. As the name implies, this is where citizens can complain about planes that fly too low over Great Britain. But according to Timothy Good's source, this was the *same* department that had an armed guard watching its door 24/7 . . . at Rudloe Manor.

That's why a number of UFO researchers, Nick Redfern among them, believe this department is actually a cover to allow the P&SS to secretly investigate UFOs, and maybe much more.

Add it all together: An inquiring government member is told that P&SS has a presence at Rudloe Manor, it being a place that sits atop an entire underground city. UFO researcher Timothy Good talks to two ex-P&SS counterintelligence agents who say you can't get at P&SS's Low-Flying Aircraft Complaints files being kept there. A UFO witness calls the top of the British government to report a UFO and is turned over to P&SS at Rudloe Manor.

5. OUR FAVORITE: THE VERY BRITISH REASON

But maybe the best clue that a British version of Area 51 exists at Rudloe Manor is a thoroughly British reason.

After all, would it not be so typically British that the section of its military that secretly handles the UFO problem would be called the Low-Flying Aircraft Complaints department?

11

UFOs over Scotland

It Takes a Village...

About thirty miles west of Edinburgh, Scotland, lies the village of Bonnybridge.

It's a typical small place for the Scottish Midlands. The population is about six thousand; industries like brick makers, saw mills and iron foundries have called it home. There's a large gravel pit on the west side of town, and Rough Castle, one of the most complete of the many Roman structures left in Great Britain, stands nearby.

But there's something very untypical about Bonnybridge: More UFO sightings have been reported here than anywhere else in the world.

The British government receives at least three hundred UFO reports from Bonnybridge in an average year. Three *thousand* sightings were reported there in just the last half of the 1990s (beating out even the noisy village of Warminster, about four hundred miles to the south). Not only has more than half of Bonnybridge's population seen a UFO; many have seen more than one.

Most of these sightings came to light in 2008 when the British government released thousands of pages of UFO reports made to the military and local law enforcement over the years. Dubbed the "Scottish X-Files," this material not only told of the many strange flying objects spotted by the citizens of Bonnybridge but also of giant UFOs landing near the town, of air traffic controllers at nearby airports tracking UFOs moving at impossible speeds on their radar screens and of airline pilots encountering bizarre flashing lights during takeoffs and landings.

Maybe it's no surprise that Bonnybridge is located inside the so-called Falkirk Triangle, a rough geometric area that stretches from Edinburgh to the towns of Stirling and Fife and back again. The Triangle is well known to UFO researchers for its many unidentified aerial sightings over the years, and the evidence seems to indicate that Bonnybridge just happens to find itself at ground zero for all this puzzling activity.

But as we will see, Bonnybridge is not the only village inside the Triangle where lots of bizarre things seem to happen.

Just like Great Britain as a whole, Scotland is no stranger to UFOs. A survey published in 2002 found that four times as many UFOs appear over Scotland per year than over places like France or Italy.

But why? And why are they concentrated over the Falkirk Triangle?

We know there's often a military element involved when a certain area experiences numerous UFO sightings—and a few military bases are located relatively close to the Triangle. RAF Tain is a weapons testing range, not unlike the Nevada Test Range, home to the Groom Lake facility in the United States. RAF Lossiemouth is a huge, bustling RAF air base located in the same general area as RAF Tain, about two hundred miles north of the Triangle. While that's a four-hour drive by car, it would take just a few minutes in a supersonic jet—or a high-speed experimental aircraft.

We spoke with researcher Andrew Hennessey, an

expert in the unusual events happening within the Falkirk Triangle. He confirms that over the last decade or so there has been a huge amount of video footage and numerous witnesses in the area recounting events that cannot be easily explained.

"The paranormal activity is ongoing as is the filming and witness reports," Hennessey said in an earlier interview. "And if the Ministry of Defence is at least partly responsible for the flying of unmanned aerial vehicles or for regular vertical takeoffs and landings from local fields by large craft—well then, folks have a right to know what or who is disturbing their peace."

There *have* been claims that the ghostly CIA-sponsored *Aurora* stealth jet has been spotted flying over the area, complete with its unique "doughnuts on a rope" contrail trailing behind it. Who knows, then, what else could be flying overhead?

Could the Falkirk UFO sightings just be top-secret military airplanes being tested inside the Triangle's airspace?

In a word, no—simply because while secret military aircraft might explain some of what people are seeing high in the skies above the Falkirk, they cannot account for the many up close and personal UFO sightings people have also witnessed in the area.

So what's going on here?

Maybe the answer lies within another of the Triangle's villages, a place that has endured events even stranger than those in Bonnybridge.

This village is called Gorebridge.

About forty miles east of Bonnybridge and fifteen miles south of Edinburgh, it is found deep within the Falkirk Triangle. With a population of about five thousand, the area is best described as rural and farmland, with a few industrial sites scattered about. Not much different in appearance from dozens of villages in the Scottish Midlands.

But while Bonnybridge has had large numbers of UFO sightings, Gorebridge can boast its fair share of UFO incidents as well.

For instance:

March 1, 2012—A strange UFO is videotaped in the night sky over Gorebridge. Spherical in shape, it's spotted hovering above the town, almost like it's watching over it, or perhaps looking for something.

March 3, 2012—The same object is seen over Gorebridge again, acting in the same manner. The only difference, this time the sphere has a weird reddish glow to it.

January 1, 2012—A distinct triangle-shaped object is caught on video by many of Gorebridge's residents. The UFO looks like the spinning lights on top of a police car, high in the sky.

November 7, 2011—A large unexplained object is videotaped flying slowly over Gorebridge. Triangle-shaped with many bright blinking lights around its outer rim, a similar UFO is spotted over nearby towns later that night.

May 2011—A video of a UFO sighting close to dusk clearly shows a round black orb flying over Gorebridge before disappearing behind a tree line.

May 21, 2009—Two UFOs intercept an airliner flying over Gorebridge on its approach to Edinburgh International Airport. The UFOs are caught on tape ascending from the Gorebridge area and buzzing the airliner from above and below via a series of fantastic maneuvers.

February 2009—Numerous reports are received of bright orange globes spotted over Gorebridge, flying in formation.

In addition to these incidents, lots of black helicopters have been reported flying above Gorebridge in recent years. There have also been reports of strange silver domes parked in nearby fields and of residents meeting people they describe as the infamous Men in Black while walking in the local woods.

Very Mysterious Midlands

But when it comes to unusual things happening in the Gorebridge area, UFOs are just the beginning of the strangeness.

LUCK OF THE ... SCOTTISH?

The area around Gorebridge is considered to be a "thin place." This is described as a location where the unusual is the usual, where the line between the possible and the impossible is fuzzy. As one example, people who live in the Gorebridge area have been documented as having better luck at winning the lottery than people living in other parts of Scotland.

CHASED BY LIGHTS

There's an abandoned coal mine in Gorebridge that's known for attracting lights in the night sky and drawing them down to it. Just as strange, one Christmas, two Gorebridge residents were cutting down fir trees near the mine when they were chased by a luminous, floating green eye.

JUMP JET ... OR ILLUSION?

There is the story of two hikers who were walking in the woods near Gorebridge when they were shocked to see a Harrier jump jet rise above a nearby tree line. But then, as they took a step back, they saw the jet go back down behind the trees. They stepped forward again and the jet rose up again. They stepped back, and the jet landed again. *What* was going on here?

THE FLYING CITY

There's another story, this from the eighteenth century, that says residents of Gorebridge once saw "a city" descending from the sky and landing nearby. While most UFO sightings in the area seem to involve normal-sized objects, historically people have seen some enormous unidentified aerial vehicles over the UK as well as other parts of the world. Could this account for the tale of a flying city?

THE MISSING LEGION

There's also a local legend concerning the fate of a large Roman army, the Ninth Legion, which was stationed in the general Gorebridge area during the Roman occupation. These soldiers simply disappeared one day, never to be seen again. Several motion pictures have been made over the years about this mystery.

GHOSTS AND WEREWOLVES

Near Crichton Castle there's a place called Valley of the Bones, where locals say a werewolf is frequently seen and heard. There are also many ghosts reportedly haunting the castle itself.

ENTER THE FAMOUS KNIGHTS

So, Gorebridge seems to be a very unusual place indeed.

But it gets even weirder. . . .

Not far from Gorebridge there's an old church called Rosslyn Chapel.

The chapel is both fascinating and puzzling. Its exterior features the standard Gothic gargoyles and flying buttresses, but inside there are some bizarre pillar carvings, a baffling ceiling design, many unexplainable "music boxes" found in the walls and more than one hundred so-called Green Men, creatures set into the masonry that are depicted with plant life coming out of their mouths.

But the strangest thing about Rosslyn Chapel is not its odd interior architecture but rather who built it and why—and what might be buried beneath it.

Though it's a topic not without controversy, there are many researchers—of both medieval history and the paranormal studies—who are convinced that this chapel, built in this very odd place, has a direct connection to the infamous Knights Templar.

A quick history of the Knights Templar will help here.

This order of warrior monks was created in France in 1118 to protect religious pilgrims traveling to the Holy Land of the Middle East. Though they started off as penniless bodyguards working for scraps, their fame and fortune grew steadily. In very little time, the Knights became immensely wealthy, politically powerful and much feared militarily throughout Europe and in the Middle East—and they stayed that way for more than 150 years.

But the Knights eventually became too powerful, so on October 13, 1307 (a Friday, making "Friday the thirteenth" a universally unlucky day), King Philip IV of France had many of the Templars arrested. Falsely charged with crimes against the crown, hundreds were tortured and burned at the stake.

However, it's believed many of the Knights were able to escape to parts unknown, taking their tremendous wealth with them.

The Big Secret

The mystery surrounding the surviving Knights and the speed with which they disappeared has led to many Templar legends.

Supposedly the Knights were in possession of some great secret that, if revealed, would have historically shocking implications. Did they know what the Holy Grail was? Did they possess the Ark of the Covenant? Did they have proof that Jesus Christ was married and had descendants?

Or was this great secret something else entirely?

Some believe that those Templars who survived King Philip's wrath traveled through a network of safe houses called *frère maisons* (meaning "brother houses"). At some point, this morphed into "Freemasons," leading to the theory that the Knights simply evolved into the Freemasons and continue to wield their power and influence today.

But others insist the Knights just disappeared—literally.

Was their great secret knowing about a place they could go where no one could ever find them? A place where they could just simply vanish?

And if so, was that place in the Scottish Midlands?

* * *

Rosslyn Chapel was built in the fifteenth century, more than 150 years after the surviving Knights disappeared. Still, some enthusiasts insist its builder, William Sinclair, was somehow involved with the Knights, pointing to what they consider to be Templar clues that Sinclair put all over the interior of Rosslyn Chapel.

For instance, only one Latin inscription can be found inside the entire church. When translated it reads, "Wine is strong, a king is stronger, females are stronger still, but truth conquers everything." This phrase was first uttered by the man who built the Temple of Solomon in Jerusalem, the *same place* where, years later, the Knights Templar supposedly discovered their big secret.

Another carving inside the chapel shows two men riding the same horse. This image is so closely associated with the Templars that it appears on their coat of arms.

Many claim that a grave on the chapel grounds marked "William Sinclair, Knight Templar" is that of the builder. This tomb also contains one of the Templars' signature eight-point crosses carved into the gravestone.

Beneath the floor of Rosslyn Chapel, though, is said to be the biggest mystery of all: a massive underground vault, sealed in 1690, that has never been reopened. No one knows what's down there, but at least one legend says a dozen Knights Templar are buried in full armor within, ready to rise again in times of need.

But the real question is, why was this odd chapel built here in the first place? Was it just a coincidence that a place of worship with such a close connection to the mysterious Knights Templar was located in an area that's now considered the world's center for UFO sightings as well as so much paranormal activity?

Inside the Chapel Walls

Whether this connection with the famous Knights is true or not, there's no denying that Rosslyn Chapel is a very curious place.

One of the pillars inside the church, called the Prentice Pillar, is ornately carved with coiled spirals that look exactly like the double helix of DNA. Is it just a fluke that hundreds of years later, just down the road at the Roslin Institute, Dolly the sheep would become the world's first DNA animal clone?

Botanists have also confirmed that there are depictions of sweet corn and cactus in the chapel masonry. The trouble is, these plants were indigenous to South

America and thus unknown in Europe at the time the chapel was built.

Many of the chapel's arches contain what people call "music boxes"—square protrusions that, by their numbers and placement, seem to present some kind of code, possibly one based on the musical scales. But if so, this code has yet to be broken.

At the top of the chapel's ceiling is the so-called Great Rose Window. At noon on the days of the summer and winter solstices, a ray of light strikes this piece of glass exactly and bathes the entire chapel in a blood-red light.

This seems like a lot of mystery and intrigue for what is basically a small country chapel.

So again, the question is why.

And why here?

Maybe one of Scotland's most famous people, Mary, Queen of Scots, said it best sometime in the mid-1500s, when she wrote in a letter to the Edinburgh city fathers, "I will always keep secret the secrets I saw at Rosslyn."

12

The False Mystery of Wenceslas Mine

Wrong from the Start

Nazi Germany was the first military power to field a jet fighter.

That airplane was the Me-262 *Swallow*. When it entered service in mid-1944, five years into World War II, it represented an impressive advance in both aeronautics and weaponry. Capable of flying more than one hundred miles per hour faster than any other airplane at that time, the Me-262, had it been built earlier and utilized better by the Nazis, might have had a greater impact on the outcome of the conflict.

But, it will be interesting to note later that at least one element of the Me-262 was decidedly *not* high-tech.

Wenceslas Mine is located in southwest Poland near the border of what is now the Czech Republic. During World War II, a German facility known as Der Riese ("The Giant") was built close to Wenceslas Mine. It's rumored that during the latter part of the war, a fantastic top-secret Nazi weapons project was housed somewhere in either Der Riese or Wenceslas Mine, making it a 1940s secret base. This project was said to be based on a machine described as being fifteen feet high, nine feet wide and made out of some kind of mysterious heavy metal. Because this device was bell-shaped, it earned the code name *Die Glocke* ("the Bell").

As the story goes, the Bell consisted of two counter-rotating cylinders with a mysterious violet liquid floating within. It is said the Bell needed tremendous amounts of electrical power to operate, yet it could be turned on for only two minutes at a time. But when it was operating, the Bell was so powerful that plants and animals intentionally positioned close to it not only died but decomposed into a sticky black goo within minutes. What's more, any human who worked around the Bell for any length of time developed sleep loss, dizziness, headaches and a metallic taste that stayed in their mouths permanently.

In fact, the residue left over from the Bell's operation was so dangerous that the Nazis refrained from cleaning it up themselves. Instead, they forced inmates from a nearby concentration camp to scour the device regularly. The aftereffects suffered by these unfortunate prisoners are unknown.

Many Things . . . and Nothing

At various times and in various media, the Bell has been reported to have been an antigravity machine, a device for creating "free energy," a power plant for Nazi UFOs, a time machine, a perpetual motion machine, a reality shifting machine, a reanimation machine and something that could manipulate time-space.

When it comes to *Die Glocke*, it's a case of multiple choice.

The story of the Bell started in 1997 when a Polish writer named Igor Witkowski claimed he'd been given access to interrogation transcripts of a former Nazi SS officer. This man was named Jakob Sporrenberg, and his only validated actions during World War II were to oversee the extermination of thousands of Jews in the

Polish ghetto of Lublin. For this, he was hanged as a war criminal. Within Sporrenberg's transcripts, Witkowski insists, were the first fantastic claims of the Bell and what it could do.

According to Witkowski, when the Bell's two cylinders began turning, with the mystery liquid (possibly mercury) within, it was said to give off high amounts of radiation. According to further writings by Witkowski and others, this counter-rotation created "vortex compression" or "magnetic field separation" or "spin polarization" or "spin resonance." Again, take your pick.

Witkowski speculates that all this was actually an experiment in antigravity propulsion. As proof, he claims that the ruins of a metal framework found in the vicinity of the Wenceslas Mine may have once served as test rig for this experiment. This structure has been fancifully dubbed the "Henge."

Nazis on the Moon?

The story of the Bell only got widespread attention in the West when retold by British author Nick Cook as well as other writers, many of whom were considered on the fringe of ufology and pseudo-scientific writing.

From this, it seems, sprang up an entire cottage industry of books, DVDs, movies and websites with fringe writers devoting much time and effort not just to the Bell, but to many more fantastic if unheard-of accomplishments of the Nazis.

Typical was the assertion that, sometime in the late 1930s or early 1940s, Germany built an interplanetary rocket base in, of all places, Antarctica. Fringe writers insisted the Nazis, along with their Japanese and Italian allies, secretly created a space program in the wilds of the South Pole and somehow achieved space flight either during or, most incredibly, *after* being roundly defeated in World War II. Some authors go on to claim the Nazis actually *landed* on the moon in 1942 and that they built bases that still exist there today. Still others claim that Hitler survived the war, is still living in Antarctica and is planning another campaign of world domination, using UFOs as his new flying panzers.

Specifically regarding the Bell, one author claimed knowledge of a meeting by the members of various secret groups—including an SS-linked group called the Elite of the Black Sun and two psychics—during which, he said, technical data for constructing a flying machine, based on propulsion technology from the Bell, came via the psychics from a planet revolving around the star Aldebaran, which is located about sixty-eight light-years from Earth.

All this makes for great science fiction—but taken as fact, these stories have become touchstones for some who'd apparently like nothing better than to keep the idea of the Nazis and their heinous policies alive.

And that's a problem.

Pure *Schwachsinn*

The simple fact is that none of these fantastic claims about the Bell are true. For many reasons.

Witkowski's original assertions were never verified. Rather, he maintained that he was allowed only to transcribe the Sporrenberg documents, as opposed to making copies and studying them deeply. Thus, these documents have never been made public. As no other facts have come to light, all the "evidence" for the story of the Bell comes only from Witkowski's claim that he saw the secret transcripts of Sporrenberg's interrogation. Not exactly scientific proof.

As for Nick Cook's book, reportedly one scientist he cites liberally is not identified by name within its pages, even though Cook claims his source is an eminent physicist at one of Britain's best-known universities. It was this blurred academic who supposedly told Cook that the Bell was actually a "torsion field generator," (whatever that is) and that the Nazis were hoping to construct a time machine with it.

Furthermore, the Henge, on which so much of the Bell's legend depends, has been dismissed as being nothing more than the remains of a conventional industrial cooling tower.

Back to the *Swallow*

Then there are more practical reasons that this is all just so much Nazi-centric nonsense, and these reasons bring us back to Germany's first high-tech jet fighter, the Me-262 *Swallow*.

At the time the Me-262 was entering the field—July 1944—the German war machine was desperate not only for oil and aviation fuel but for basics like steel, aluminum and rubber. That fact is, Germany was so strapped for metallic resources in 1944 that it was forced to use plywood inside the cockpit of the Me-262 jet fighter. In addition, several other "wonder" aircraft, all with more potential than the Me-262, were basically left on the drawing boards because of Germany's lack of raw materials.

Where, then, would the Nazis get the resources to study, design and build an antigravity machine or a time machine or a fleet of UFOs—or for that matter, a space base in Antarctica from which they could reach the moon?

Where was the huge infrastructure that would be needed to produce the technology to do these fantastic things? Even if the Nazis had theoretically solved the myriad technological challenges of space travel, to reach the moon in 1942 would have required, at the very least, an effort along the lines of the Manhattan Project, America's massive development program to produce the atomic bomb. That program, which cost in today's dollars more than $25 billion, utilized more than 130,000 people in nineteen different locations scattered across the United States, and it took six years to complete.

Why was no evidence of a similar concentrated effort found once the Allies entered Germany at the end of the war?

Numbers Confirm the Lie

Further proof that this is all hokum comes from the stories about what happened to the Bell after the war.

Witkowski speculated it ended up in a country friendly to the Nazis. Others claimed it was moved to the United States along with the dozens of German scientists who were allowed to come to America after the war. More claims say when the war was drawing to its conclusion and the Nazis were being steamrolled from all sides, the Bell was put on a long-range aircraft, already in short supply in the Nazi Reich, and flown to . . . Argentina. Still others believe the Bell was transported by U-boat to the secret ethereal Nazi base in Antarctica.

But again, none of this could be true. If the Bell was an antigravity device, and was transported to the United States, this would mean American scientists would have had it in their possession now for nearly seventy years. Where is it then? Why haven't we seen any fruits from these seven decades of research? And why would the United States go through with a charade like NASA and spend billions of dollars on massive and highly dangerous liquid-fuel-propelled rockets at

the cost of eighteen astronauts' lives if they had in their
hands a working antigravity machine?

As for the Bell being flown to Argentina in a
multiengine German aircraft in May of 1945—simply
put, such a trip would have been impossible without
refueling. At the time, Germany's largest and heaviest
airplane was the Ju-390. Its range was about six thousand miles. From Poland to Argentina, under the best
of conditions, is an eight-thousand-mile, twenty-five-
hour flight.

Where between Poland and Argentina could a Nazi
aircraft stop and safely refuel? Or perhaps the better
question is, who would refuel a Nazi aircraft anywhere
in the spring of 1945?

The same is true for a trip to Antarctica by a U-boat.
At the end of the war, the Germans' farthest-ranging
submarine was the so-called Type IX. True, it could sail
eight thousand miles without need for refueling, but
that was its range while traveling on the surface of the
ocean, not beneath it. This means any Type IX U-boat
would have been forced to make the entire trek to
Antarctica in full view of flocks of Allied aircraft patrolling the Atlantic by 1945 and would have been spotted
easily. Besides, the distance between Germany and Antarctica is more like ten thousand miles. Where would

this Bell-carrying U-boat have refueled to make such an arduous monthlong trip?

And finally, if the Nazis had such fantastic technology as an antigravity machine, a device for creating free energy, a fleet of UFOs, a time machine, a perpetual motion machine, a reality-shifting machine, a reanimation machine and a time-space manipulation machine, plus space travel and bases on the moon, then why did they lose the war?

Was It Anything at All?

If it did exist, from the bare description, the Bell might have been a crude heavy particle accelerator, something the Nazis used in a desperate attempt to breed bomb-grade uranium in order to make an atomic weapon. But again, the Nazis had nowhere near the resources the United States had in the same kind of quest, so luckily they were doomed to fail.

But one thing from those fantastic stories *does* have an unintentional bit of truth to it. It has to do with a landing on the moon.

At the end of the war, the United States did indeed allow many Nazi scientists to enter the country and work on America's then fledgling space program. This questionable immigration was called "Operation Paper-clip," and the argument of its morality goes on too long to be hashed over here.

But this is for certain: These Nazi scientists, the most famous among them being Wernher von Braun, formed the foundation of NASA. They designed and built the Saturn V, the blockbuster rocket that had enough lifting power to boost the Apollo lander into orbit and beyond.

In other words, the Nazis *did* put someone on the moon: the United States of America.

13

The Case of Saddam's Area 51

Secrets in the Desert

A crashed UFO in the desert. Strange debris seen by witnesses. The U.S. military gets involved, claims the wreckage and then clamps down on all media coverage.

Sound familiar?

But this isn't Roswell, New Mexico, in 1947. This incident, some say, happened in Saudi Arabia in 1991, during the first Persian Gulf War.

It all began one night early in the conflict. A flight of four U.S. F-16 warplanes heading for a mission over Baghdad was alerted by Saudi air control that an unknown aircraft was heading in its direction. This unknown aircraft turned out to be a UFO.

One of the F-16s peeled off from the formation and began chasing the UFO. The UFO changed directions in an attempt to escape, but the F-16 continued the pursuit. The UFO apparently fired a weapon at the F-16 but missed. The F-16 returned fire with two Sidewinder missiles—and both hit the saucer-shaped craft. The UFO blew up, its wreckage going down in flames.

A Russian military attaché was in the Saudi air control center when the incident occurred; he was able to determine where the debris came down. Quickly securing transportation, he was among the first witnesses to reach the desert crash site.

The Russian reported seeing small seats and instruments inside the heavily damaged saucer, a description similar to what Bob Lazar claimed he saw inside the saucers supposedly kept at S4. But then the U.S. military arrived on the scene and escorted the Russian and his colleagues away. Later on, the Russian learned the debris had been packed up by the U.S. Army and shipped off to parts unknown.

It's a strange story—but as it turns out, it's not the strangest UFO tale to come out of that troubled part of the world in recent times.

In fact, some think that another crashed UFO incident, this one inside Iraq itself, led directly to the U.S. invasion of that country in 2003.

Mission Accomplished?

On March 20, 2003, the United States led a coalition of approximately two hundred thousand troops in invading Iraq.

According to U.S. President George W. Bush at the time, the main reason for the invasion was to disarm Iraq of weapons of mass destruction.

After three weeks of heavy fighting, this United States succeeded in seizing the Iraqi capital of Baghdad and toppling the government of Saddam Hussein.

On May 1, 2003, Bush landed on the aircraft carrier USS *Abraham Lincoln* and gave a speech announcing the end of major combat operations in Iraq. Clearly visible in the background was a banner famously stating "Mission Accomplished."

That banner and the speech were criticized at the time for being premature—and history bears this out. More than 4,500 U.S. combat troops would die, and tens of thousands would be wounded over the next several years, fighting a bloody battle against an Iraqi insurgency that rose up as a direct result of the invasion. Meanwhile, several media outlets reported that at least one hundred thousand Iraqi civilians died as a result of the war, possibly many more.

There are some who believe George W. Bush came to office in 2001 with plans already in place to invade Iraq, the aim being to gain control of that desert country's vast oil supplies. Indeed, at first the U.S. military operations were conducted under the code name Operation Iraqi Liberation (OIL). This was later changed to Operation Iraqi Freedom.

Bush's detractors claim further that his stated goal of seeking to disarm Iraq of weapons of mass destruction was just a ruse. And in fact, in 2005 the CIA released a report saying that no nuclear, biological or chemical weapons were ever found in post-invasion Iraq.

But if no nuclear, biological or chemical weapons were found, then what was Bush talking about when he announced "Mission Accomplished"?

Wasn't finding these kinds of weapons what the war was all about?

Or was there another reason?

Weird Doings Before the Start

There's no doubt odd things were going on inside Iraq just before the start of the war.

The hostilities actually commenced on March 19, 2003, with two spectacular, well-publicized air strikes:

one on a suspected hiding place near Baghdad where Saddam Hussein was mistakenly thought to be (this was the first and most famous "shock and awe" bombing) and the other on the city of Mosul.

However, eight hours before any of that happened, some sort of battle took place over the Little Zab River Valley, about 175 miles north of Baghdad, near the city of Zarzi—a battle that was precipitated by the appearance of strange lights in the sky.

On that night, a small army of U.S.-allied Kurdish militia was in control of some high ground just above the Little Zab River Valley. While awaiting the opening shots of the war, these militiamen saw a group of unusual lights flashing all over the sky.

The Kurdish soldiers were confused. What were these lights? Some thought they were lightning bolts from a storm on the western horizon that was rapidly moving east. Others thought the flashes might be the opening shots they'd been waiting for, just coming early. Were these smart bombs or cruise missiles going off somewhere in the distance?

No one knew. But at some point, Iraqi antiaircraft batteries about five miles away from the Kurdish position began firing at the lights. This only added to the confusion. Did Saddam's AA crews think the lights belonged to Coalition airplanes? Or a fleet of spy drones, perhaps?

Time passed and then more strange lights were seen. For some reason the Iraqis started shooting off flares and then streams of tracer fire. While all this was seen clearly by the Kurds, they could not determine what the Iraqis were shooting at over the Little Zab River.

Not too long after, the war began for real.

The Mystery at Qalaat-E-Julundi

As the war raged, it was apparent to some UFO researchers that American forces were paying special attention to the Little Zab River Valley, particularly the same area where the strange lights had been seen hours before the war started.

U.S. aircraft bombed targets in the region throughout the conflict. Reports say thousands of munitions, some dropped by massive U.S. Air Force B-52s, poured out of the sky on many occasions, all targeted on places in the Little Zab River Valley. Even in a country lit up with war from one end to the other, these actions near Zarzi seemed particularly vicious and intense.

Why here?

Because, some say, this part of the Little Zab River Valley is where Qalaat-e-Julundi is located. And Qalaat-e-Julundi is where Iraq's version of Area 51 could be found, a place where some claim Saddam had some very exotic and very deadly items hidden away—possibly including a crashed UFO.

Why Would Saddam Need His Own Area 51?

At one time, Iraq had one of the largest militaries in the world. Supplied mostly by Russian and French arms manufacturers, by the early 1990s, Saddam's forces boasted hundreds of modern tanks and jet aircraft as well as hundreds of thousands of well-equipped troops.

At the very least, such a large standing army would require a place to work on its own specialized weapons or weapons adaptations. In Iraq's case, for instance, things needed for the specifics of desert warfare.

Plus it was well known that at one time Saddam did indeed have an extensive arsenal of poison gas. He'd used this gas on his own people in 1988, killing as many as five thousand innocent Kurds in northern Iraq. Such WMDs were probably developed and weaponized at Qalaat-e-Julundi.

There are also whispers that Saddam had an even more powerful weapon at his Area 51, something that might have led directly to the destruction of the space shuttle *Columbia*. That ill-fated spaceship blew up as it was entering Earth's atmosphere on February 1, 2003, shortly before the invasion of Iraq began. On board was Colonel Ilan Ramon, Israel's first astronaut and one of the pilots who, nearly twenty years before, bombed Iraq's Osiraq nuclear reactor. There was some speculation that the reason Ramon was part of the shuttle crew in the first place had something to do with gathering intelligence on Iraq every time the spacecraft passed overhead.

Because the Iraqis probably suspected this orbital spying, and because one of their most hated Israeli foes was on board, some believe an Iraqi laser weapon of some sort may have fired at the shuttle on its last pass, inflicting enough damage to cause it to break apart and crash.

Unlikely? Maybe. But if true, such a weapon would probably have been developed at Qalaat-e-Julundi, too.

Whose WMDs Were They?

Maybe such tales are the reason why another fantastic story about Qalaat-e-Julundi has shown surprising resiliency: that Saddam had a crashed UFO, aliens and possibly some kind of horrific weapons hidden away at the Iraqi Area 51.

This particular scenario starts back on December 16, 1998, when a video clip, said to have aired on CNN, appeared to show a UFO flying over Baghdad. The object was seen trying to avoid antiaircraft fire being directed at it by Saddam's forces.

At the time it was assumed this was just another UFO sighting, one of hundreds that take place around the world every day. But then rumors started bubbling up that maybe the UFO was actually *hit* by the AA fire, shot down and recovered by Saddam's military.

As the story goes, the wrecked UFO and possibly its surviving occupants were taken to Qalaat-e-Julundi. Once there, Saddam had his weapons experts reverse-engineer one aspect of the technology found on board, and from *that* they created some really mind-boggling WMDs.

So, was it *this* fantastic weapons cache that the United States was secretly looking for all along? Could some elusive extraterrestrial-based WMDs have been the reason for the invasion in the first place?

Bizarre Bazaar

Though it sounds outrageous, some scientists inside Iraq didn't totally discount the story.

One scientist was quoted before the 2003 invasion as saying, "It is rumored at a market in Sulaimaniya, to the south of Zarzi, that aliens are Saddam's guests. Where do they stay? People mention some underground base at the old stronghold Qalaat-e-Julundi. It is practically impossible to penetrate it. The citadel stands on a hill surrounded with vertical precipices on three sides and these precipices plunge down to the Little Zab River."

It's possible, though, this tale was created solely to keep the highly superstitious Iraqi people away from Qalaat-e-Julundi. Rumors are a big part of Iraqi life, usually the more outlandish the better. For instance, after the 2003 invasion, many Iraqis were convinced that the wraparound sunglasses worn by many U.S. soldiers either contained X-ray vision or projected maps showing the location of every house inside the war-torn country.

Bottom line, it's hard to tell fact from fiction in the bazaar.

The Russian Version

What's interesting about this Iraqi Area 51 saga is that much of it came from the same Russian colonel who'd reported the dramatic F-16/UFO shoot-down story from the first Gulf War, a dozen years before. His account of the extraterrestrial goings-on at Qalaat-e-Julundi appeared in *Pravda*, the onetime official newspaper of the Soviet Union. And while the *Pravda* of the old Moscow regime doesn't exist now as it did then, it remains a media outlet in Russia. A bit sensationalistic, but still highly read.

As one Russian journalist explained to us, "*Pravda* today is sort of 'yellowish.' They don't have any correspondents' network or stringers. They are sensation hunters and usually compile secondhand news from the Internet.

"However, that does not mean what they write about is not true."

Legacy TBD

So was the massive bombing around Zarzi and the Little Zab River Valley a preparation for an assault on Qalaat-e-Julundi itself?

Once the area had been pummeled, did U.S. troops secretly attack Saddam's Area 51 and actually take possession of those alien WMDs? Not nuclear, not biological, not chemical—but something still very dangerous that had an unearthly origin?

Could *that* have been what "Mission Accomplished" was all about, putting to bed the speculation that Bush would be so deceitful that he would send U.S. forces to war under false pretenses? There's no disputing that Bush's two terms were bookended by near catastrophic events for the American public, from the September 11 attacks in 2001 to the financial crash of 2008. But did he *really* order the 2003 invasion of Iraq because of some reverse-engineered alien WMDs?

We may never know.

But considering how Bush got elected in the first place, by a hanging chad, and how he almost died from eating a pretzel while in office, and that his ranch was extremely close to one of the largest and most mysterious UFO incidents in U.S. history—the famous July 2008 Stephenville sighting—then anything is possible.

14

The M-Triangle

Heaven on Earth?

The vast underground base at Yamantau is not the only mysterious place in Russia.

The former Soviet Union boasts a number of unusual locations—including places where UFOs are frequently seen and either welcomed or shot at.

One of the most curious in the former group is the so-called M-Triangle; there might not be another place like it anywhere.

So many highly unusual things go on there that some say the M-Triangle might literally be a piece of Heaven that somehow wound up here on Earth.

* * *

The M-Triangle is located about six hundred miles east of Moscow in a remote region of Russia near the Ural Mountains.

Also known as the Perm Anomalous Zone, the Triangle encompasses about forty square miles of mostly mountainous terrain near the villages of Molebka and Kamenka, close to the Silva River.

According to the locals, strange things have been happening in this area for hundreds of years. Bizarre lights in the woods, unidentified flying objects streaking overhead, frequent encounters with otherworldly beings, even weird symbols and letters written across the sky. Something highly unusual is said to occur inside the zone almost every day, and reportedly, many times, dozens of different weird things are going on there at once.

As is almost always the case, the M-Triangle's strangeness begins with UFOs. They usually appear over the zone as brightly lit objects. Local villagers say they come in many colors and shapes, including spheres, domes and saucers. Sometimes these objects just pass overhead; other times, they hover above the zone for hours.

These brightly lit objects have also been seen in the nearby woods themselves—floating low to the ground, moving in and out of the trees like phantoms. So many UFOs are spotted around the Perm Anomalous Zone that a large wooden statue has been erected at the entrance to the village of Molebka to commemorate these frequent otherworldly visits.

It Gets Even Odder. . . .

But again, researchers say other bizarre things happen inside the M-Triangle, too, including the following:

SOUND MIRAGES

Sound mirages can be several different things. One frequent sound mirage is a real sound that apparently originates from far away yet can be easily heard within the Perm Anomalous Zone. For instance, one research group, deep in the woods within the M-Triangle, heard the sound of a car coming toward them, an impossibility due to the thickness of the forest. Another group of researchers spending the night inside the zone heard strange voices from dusk to dawn, voices so clear and distinct they seemed just inches away. A ghostly whistle, an electrical buzzing noise and the sound of ancient choral singing are frequently reported inside the zone as well.

STRANGE FIRE

Fire behaves very peculiarly inside the M-Triangle. Researchers have reported many instances in which a flame of any size will suddenly explode, as if being fed by an invisible propellant like gasoline. Scientists have no clue why this occurs.

THE CALL BOX

There is virtually no cell phone service inside the M-Triangle—except at one place researchers have dubbed the "Call Box." They say that if a person stands inside this five-foot-by-five-foot-square piece of ground, for some unknown reason they can make a cell phone call to anywhere in the world.

FROZEN TIME

Possibly the first to report the frozen time phenomenon was a group of researchers who went into the M-Triangle in 1990. Several times during their foray all the members' electronic watches stopped at precisely the same moment and began flashing "00:00."

DIFFERENT TIME

Researchers say they are also baffled why in some places within the zone watches show a time that is hours ahead of the actual time, whereas in other places watches run hours behind.

THE TREES HAVE EYES

Many researchers who have gone into the M-Triangle report it's impossible to shake the feeling that someone or something was watching them at all times.

Tearing Down the Wall

The strange events inside the M-Triangle weren't very well known until the late 1980s.

Up to then, the old hard-line Soviet government had forbidden journalists from reporting or writing on the subject of UFOs anywhere in the USSR, and that included rumors that something strange was happening in the Molebka section of the Urals.

As one Russian researcher told us in an interview, back then, "Any talks on the matter might get you in trouble."

This also meant physical access to the M-Triangle

was nearly impossible. Even though the old Soviet government sent several expeditions into the region over the years (what they found was never disclosed), the area was strictly off-limits to civilians, and most especially to UFO researchers.

When the crumbling Soviet government finally relented and removed their ban on UFO reporting in the late 1980s, many researchers jumped at the chance to study what was occurring inside the M-Triangle. A few civilian expeditions finally went in, and their results caused a sensation. This information reached the world media in 1989 and research groups have been studying the Perm Anomalous Zone ever since.

But Russia being Russia, even these days researchers claim the entire area is under constant surveillance by the Kremlin's intelligence agencies.

The Stars Go Dancing

Valery Yakimov, director of the Ural-UFO Group, reports extensive experiences inside the M-Triangle.

He recalls that, while spending time there as a child, there were parts of the M-Triangle that adults warned children to avoid. As Yakimov grew older he began seeing UFOs, of varying shapes and sizes, over the zone.

Yakimov returned to the M-Triangle in 1989, as an adult. His excursions produced fantastic observations.

One night, Yakimov and his group saw what they described as an unusually shaped sky. The stars above became so concentrated they formed a dome. Other parts of the sky appeared normal.

On another occasion, Yakimov's group saw around two dozen stars moving in inconceivable trajectories. They floated over the researchers' heads and formed a gigantic circle. This vision was seen for four nights, between midnight and one a.m.

During one expedition, the group observed a number of strange geometrical shapes—squares and triangles mostly—floating through the forest. They showed up at the same time, at the same location, and they could be seen clearly with binoculars.

Varying from silver to white to blue, they appeared one at a time and remained visible until someone tried to get closer to them, at which point they disappeared. Though dozens of people saw them, the shapes couldn't be captured on video.

During another expedition, Yakimov's team saw many unexplained spheres gliding over the area. White or orange in color, they had diameters of three to ten feet. Sometimes singly, sometimes in groups, the spheres approached the team and floated over their heads.

The members got the impression the spheres contained intelligence, because as they sailed over they sometimes reduced their brightness, or even stopped occasionally to hover over the group. It was as if they were observing Yakimov's crew, just as the researchers were observing them.

A similar occurrence was related to us in an interview with M-Triangle expert Nikolay Subbotin, director of the Russian UFO Research Station (RUFORS). Camped out for the night deep inside the zone, Subbotin and another group of explorers spotted a brilliant ray of light coming from behind a grove of trees and rising up to the sky. The group froze, stunned by the unexplained illumination. An instant later, a second ray appeared, parallel to the first. Both slowly began to move in opposite directions, lighting up the area all around them.

"It was a fantastic display and totally incomprehensible," Subbotin told us.

Subbotin and his friends watched this extravaganza for about fifteen minutes before being distracted by another incredible display. The group realized that the clouds over their heads had begun to line up in an "absolutely pristine lattice formation." Formerly cumulus and puffy, the clouds suddenly consisted of five vertical and five horizontal stripes positioned in perfect rows.

Subbotin reports the clouds stayed like this for a half hour before finally "melting" and disappearing for good.

"The Zone Is Activated"

Other people entering the Perm Anomalous Zone report events more down-to-earth, though no less amazing.

Another story related to us by Nikolay Subbotin tells of a husband and wife team visiting the zone for the first time. Early one morning, while preparing to float down one of the area's rivers in an inflatable boat, they became aware of a strange buzzing noise all around them. They described it as sounding like a water pump, working away.

They began their journey downstream, but the noise never left them. They heard it for hours, no matter where they went, no matter what direction they were heading. This was the case for their entire visit. They joked that the noise meant "the zone is activated."

The couple returned to the zone for a second time later that year and the same noise was still audible. Yet on a third trip it was absent.

To them, this meant the zone was "turned off."

The same couple noticed odd things about the zone's animal life as well. While wading in shallow waters of a river, they discovered fish had no qualms about swimming right up to them and eating food out of their hands. Butterflies and dragonflies would also frequently land on them, and stay in place, totally unafraid of human contact.

MEET THE ALIENS

But apparently there are other creatures roaming around the Perm Anomalous Zone, ones this world is not so familiar with.

Glowing ETs and other nonhumans have been frequently reported in the M-Triangle. Some of the latter, known locally as "Snow People," fit the description of Bigfoot to a T. Others have even more bizarre appearances.

Researcher Valery Yakimov says his most chilling M-Triangle encounter involved meeting one of these alien beings face-to-face.

It happened during an expedition in which Yakimov had brought some people into the zone who had never been there before.

Yakimov took the rookies to a spot where researchers say they usually feel a presence of "the unknown." At that point, it was suggested that the group split up and walk about the zone one at a time. This way the new-comers would be able to experience the full intensity of the M-Triangle.

This was why Yakimov was walking alone when he says he was confronted by a creature almost nine feet tall.

"I came down a small ravine," Yakimov wrote soon afterward, "and suddenly something made me look up. I saw a being of nine feet tall. He was black in color, nontransparent, proportional. I was horror-stricken."

Frozen to the spot for a few moments, Yakimov was finally able to get his feet moving. He ran . . . and ran . . . and ran, eventually finding another member of his group.

When Yakimov told the fellow researcher what had just happened, his colleague was understandably aston-ished. They quickly returned to the spot where the rest of the group was supposed to reunite.

Although Yakimov did not mention his encounter to anyone else in the group, a short time later one of the newcomers, a young woman Yakimov did not know, sought him out. She told him that she'd had a meeting with a strange being as well, and at a place near Yaki-mov's encounter.

She had been walking alone through a meadow when she was suddenly face-to-face with an unearthly being. While Yakimov had seen a creature more than nine feet tall, the female observer had encountered what she described as a little green man, about three feet tall. This being walked along with her for a while and then touched her, burning her skin slightly. That's when the woman turned and fled, eventually returning to the rest of the research group.

Later on, Yakimov and the woman went back to the area where they had seen the strange creatures. Yakimov reported that they were not frightened anymore. They both felt a sense of overwhelming peace and well-being, something reported by just about everybody who visits the zone.

In another story relayed to us by Nikolay Subbotin, a researcher camping out in the zone went to collect some water. Walking back to camp, he realized someone was following him. He immediately began to run, but not before getting a good look at his stalker, which sounded close in description to the being Yakimov had encountered.

"There was no bend in the [creature's] wrists," Subbotin said, recalling the researcher's encounter. "The being walked slowly. This creature was black, with white stuff on its body. It wore a kind of skintight skirt and was about nine feet tall. The eyes were large and shiny and it had long arms."

Why Is the Zone the Way It Is?

Nikolay Subbotin told us there are a number of theories why the M-Triangle is such a *drugoe mesto*, a "different place."

First, there are rumors that a secret weapons facility is located somewhere under the Perm Anomalous Zone. This would account for the constant electrical sounds heard in the area, as well as reports of people meeting members of the Russian military while exploring the M-Triangle. It would also account for why the place is of such interest to Russian intelligence agencies.

There are also stories about past attempts at mining strontium in the area. Strontium is an earth metal that's used to block X-rays in color TV sets, among other things.

Other UFO researchers theorize that uranium found in the nearby mountains might be attracting UFOs. Some have even speculated the presence of this uranium might be causing mass hallucinations within the zone.

Subbotin told us of another piece of information

he'd picked up: He'd heard that in the 1950s and 1960s, a highly classified underground test laboratory inside the zone had conducted secret experiments into large electromagnetic fields. Supposedly something went wrong in the course of these experiments, and as a result everyone involved had to be evacuated from the area. The underground laboratory was then flooded as a safety measure.

But Subbotin is quick to remind us that all these theories are mere speculation. He says that no solid reason has ever been given as to why things are the way they are inside the M-Triangle or why local villagers say strange things have been happening in the zone for centuries.

A Piece of Heaven

But what *really* makes the M-Triangle different is the astonishing personal effect it has on many of the people who spend time there.

While there have been stories of some people being adversely affected within the Perm Anomalous Zone—burns, confusion, even reports of a suicide—more often the tales are about how a person's well-being becomes vastly improved after a trip to the M-Triangle.

For instance, there have been reports of people suffering from serious medical ailments visiting the M-Triangle and coming out completely cured. Moreover, many healthy people who venture into the zone leave feeling what one Russian researcher called a general improvement in all spheres. Not only do people feel better physically and mentally after being inside the M-Triangle, but they also feel different *spiritually* and *morally*.

The researcher continues: "The drawbacks of one's character seem to disappear here; the good intentions and high feelings come to life. One's soul becomes cleaner, higher and calmer. Nearly everyone feels a unique comfort of soul."

In other words, they come out an all-round better person, which is why many people feel compelled after their first trip to the zone to go back again.

From the Zone to the Stars

There's also a kind of creative effect that reportedly infuses some people who spend time in the Perm Anomalous Zone. Abilities become sharper, previously unknown talents come to the fore.

There is one story in particular that would seem to

underscore to even the most skeptical person that something extremely unusual is happening inside the M-Triangle.

It is the story of Pavel Mukhortov. Discharged from the Soviet Army in the 1980s for health reasons, and for a time unemployed, Mukhortov eventually found work as a journalist. He'd heard about strange things going on inside the M-Triangle and considered visiting the place to see for himself, but he wasn't sure. But when he learned the KGB was continuously monitoring the area, Mukhortov found the impetus he needed. He joined a research party and set off to explore the zone.

Immediately on arrival, Mukhortov sensed something eerie about in the place. While some members of his group soon fell sick, personally Mukhortov found himself enveloped by uncontrollable emotions.

Photos of strange flying objects were taken during this expedition, and according to a *People* magazine article from October 1989, Mukhortov claimed to have met and spoken with an alien while inside the M-Triangle.

But the *really* strange stuff started happening once Mukhortov left the zone and returned to Moscow. Soon after arriving in the Russian capital, the reporter began to feel radically different. Hardly a genius when he went on his trip to the zone, he suddenly found himself saturated with knowledge about aerospace physics—something he'd been completely unfamiliar with previously.

Incredibly, he'd become *so* enlightened in this area that he eventually applied for and was accepted to the Russian space program.

He became a cosmonaut a short time later.

15

Kapustin Yar

The Opposite of Heaven

Cosmonaut savant Pavel Mukhortov probably became familiar with another place that, while located eight hundred miles southwest of the M-Triangle, was light-years away on the cosmic enlightenment scale.

This place is called Kapustin Yar. Found in the desert of northern Astrakhan Oblast near the city of Volgograd, it is, primarily at least, a massive rocket launch and development facility, part of both Russia's never-ending drive for weapons research as well as its long-enduring space program.

But if even one of the more fantastic stories coming out of this place can be believed, Kapustin Yar has also been the site of nothing less than a secret war between the Russians and whoever, or *whatever*, flies UFOs.

Unlucky Nazis

The history of Kapustin Yar begins with Nazi scientists.

In the final days of World War II, many of the German rocket engineers who'd helped Hitler devise his so-called wonder weapons such as the V-1 buzz bomb and the larger V-2 missile found themselves caught between two worlds. With the Third Reich collapsing, many of these scientists fled west and enthusiastically surrendered to the U.S. Army. As mentioned earlier, under a secret program called Operation Paperclip, they were brought to the United States and became the foundation of what we now call NASA.

Those German scientists who weren't so lucky, though, were captured by Soviet troops advancing from the east. Forced to work on Russia's then fledgling space program, many of them wound up at Kapustin Yar.

The Secret City

While these days it's found close to the western border of Kazakhstan, back in 1946, the year it was built, Kapustin Yar was located deep inside Soviet territory.

An entire city, named Znamensk, was built to house the thousands of people the Soviets assigned to this facility. For a long time, Znamensk was a secret city; it did not appear on any maps and was off-limits to outsiders. In fact, Kapustin Yar was deemed *so* secret that the nearby small town of Zhitkur was considered too close to it. So the Soviets emptied out its population and razed it to the ground.

The United States and its allies only became aware of all this when one of the captured German scientists who'd been working at the secret missile base somehow made it back to Germany and was interrogated by Western intelligence. A spy flight went over the area in the summer of 1953 and confirmed it: launchpads, highly sophisticated radar and at least one ultra-long runway.

Kapustin Yar was for real.

Everything from orbital spacecraft to nuclear-tipped ICBM test missiles to components of Russia's ultimately failed version of a Star Wars–type antimissile shield—

more than 3,500 major launches in all—have blasted off from Kapustin Yar since it first opened almost seventy years ago.

Dark Skies Above

But Kapustin Yar isn't just the Russian equivalent of Cape Canaveral. It's actually more of a combination of Cape Canaveral and Area 51.

Like its American counterpart, Kapustin Yar boasts numerous "unidentified areas," buildings whose purposes can only be guessed at.

Also like Area 51, Kapustin Yar has a long history of UFO incidents.

But unlike the Perm Zone, this UFO connection does not involve human enlightenment or visitors coming away with a higher sense of intelligence or spirituality.

Just the opposite.

Russian UFO enthusiasts claim the first extraterrestrial event happened at Kapustin Yar in 1948 and it was anything but benevolent. While guiding a Soviet fighter plane in for landing, the base's radar station reportedly picked up a UFO flying close by. Quickly alerting the plane by radio, its pilot was said to have spotted the strange object as well. He described it as being silver in color and shaped like a cigar.

As the story goes, word of what was happening was flashed directly to Josef Stalin, then dictator of the Soviet Union. Probably believing the object was a spy plane sent from the West, Stalin ordered the fighter plane to shoot it down.

But it was not as simple as that. While the Russian pilot attacked as ordered, a prolonged dogfight ensued. The UFO and the Russian airplane began firing their respective weapons and reportedly both sides scored what turned out to be mortal blows. Both craft eventually crashed. While the fate of the Soviet pilot was never revealed, multiple sources say the UFO's debris (possibly along with the bodies of its occupants) was recovered and brought to Kapustin Yar.

Thus began what some claim would become Kapustin Yar's long and secret war with UFOs.

War of the Worlds

Reports say this secret war persisted during the 1950s and 1960s, with UFOs regularly showing up over Kapustin Yar and Soviet fighters regularly scrambling to intercept them. Air battles were frequently the result, and in retaliation for these clashes the UFOs would sometimes prevent missiles on Kapustin Yar's launch-pads from taking off.

One particularly detailed incident, reported on both Nicap.org and UFOCasebook.com, was said to have occurred in August of 1989, when Soviet MiG fighters intercepted and shot down a UFO near the city of Prokhladny in the North Caucasus region of the old Soviet Union, an area close to Kapustin Yar.

The remains of the UFO were eventually located and a Soviet recovery team was dispatched to the scene. The mysterious object was said to be twenty feet long, ten feet high and shaped like a cigar. Some reports say three occupants, each about three feet tall, were found inside the object when the Soviets arrived. Two were dead and the third died shortly afterwards.

Eventually the recovery team was able to retrieve the crashed object by using a heavy-lift helicopter. While the remains of the three occupants were taken to a secret location, the object itself was said to have been flown to Kapustin Yar.

According to some reports, this was just one of at least a handful of crashed UFOs that have been brought to the secret missile base over the years to be examined and then locked away.

An Outspoken Hero

There's no question these stories sound incredible. Dogfights? Crashed UFOs? A secret war with ETs?

But here's where the story of Kapustin Yar takes an even stranger twist.

In just about all the instances mentioned above, among the people reporting them is a person held in very high esteem in aeronautical circles in both Russia and the West.

Her name is Marina Popovich. A legendary Russian test pilot, Popovich, who was born in 1931, owns dozens of the world's top aviation records. Very well known in Russia, she's been awarded the Hero of Socialist Labor medal, the Order of Courage and has a star named after her in the constellation of Cancer.

Popovich has written a number of books, including one titled *UFO Glasnost*. Published in 2003, the book claims that Russian pilots, both military and civilian, have made more than three thousand *confirmed* UFO sightings over the years. Popovich also says Russian intelligence services have in their possession fragments of at least five crashed UFOs, debris that Russian scientists have concluded is simply not of this Earth.

More ominous, though, Popovich has also said she's

personally witnessed aerial combat between Russian fighter planes and UFOs.

One such incident happened in 1964. Aloft during a routine training mission, a group of Soviet jets encountered a UFO. In the confusion that followed, the UFO fired some kind of weapon at them. The Soviet jets immediately took evasive action and prepared for combat, but the UFO suddenly disappeared and no further shots were fired.

Another incident Popovich reported happened on August 7, 1967. A Russian pilot encountered a UFO that suddenly projected some kind of light beam in his direction. Despite the pilot's best efforts to avoid it, one of his wings came into contact with this beam, causing the pilot to lose control of his aircraft. With the plane shaking wildly and his instruments going haywire, the pilot briefly glanced outside to see that the affected wing was glowing brightly. He somehow managed to land the plane, touching down even as the wing was still glowing. In fact, some sources say the wing continued to glow for weeks afterward.

Popovich has also said her fighter squadron was regularly scrambled to do battle with UFOs specifically violating the restricted airspace above Kapustin Yar.

This secret war went on for so long and was so intense, Popovich says eventually the two sides just settled on a sort of unspoken truce. The Russians stopped firing at the UFOs over Kapustin Yar and the UFOs stopped interfering with the Russian missile launches below.

(None of this was lost on the KGB, of course. By the mid-sixties, it had opened up its own secret study into UFOs. Known as the Blue File, the study contained more than a hundred pages of UFO incidents over Soviet territory; Popovich has called it the largest study of UFOs ever commissioned by any government organization, anywhere in the world.)

Just how highly is Popovich regarded in Russia? Her popularity has been compared to that of both U.S. astronaut Neil Armstrong, the first man to walk on the moon, and Chuck Yeager, America's most famous test pilot.

And again, while her stories all sound too far-fetched to be true, why would someone in her high-ranking position lie about such things?

Which leads to another question: Would people be more likely to believe another story about UFOs at Kapustin Yar if the person vouching for the story had a famous American name such as . . . Rockefeller?

A Chilling Incursion

Not all the UFO incursions around Kapustin Yar were confined to the air above the secret missile base.

In 1989, there was a major incident at a weapons installation located within the launch center itself. A partial file of this event was declassified by the Russian intelligence service in 1991; it contained depositions from Soviet military personnel, both officers and enlisted men, who witnessed the incident firsthand. It also contained illustrations of the object drawn by some of the witnesses, plus a summary written by the KGB. While there is no official conclusion about the event, the file is fascinating in the detail it contains.

Investigative reporter J. Antonio Huneeus wrote an intriguing account of this alarming incident in a document titled "Unidentified Flying Objects Briefing Document: The Best Available Evidence" (more on this below).

It says that around midnight on July 28, 1989, a UFO appeared above the Kapustin Yar base and was seen by several members of the Soviet military. The KGB later questioned these men and their recollections were similar in content.

One of those questioned, Ensign Valery Voloshin,

described the UFO as a disc-shaped object some seventeen feet in diameter. He said its hull had a dull green glow and that a powerful light on its underside blinked like a camera flash. The UFO flew very low over various buildings on the Kapustin Yar base before heading toward the installation's weapons depot.

Once over the depot, a place where nuclear warheads were kept, the UFO went into a hover and the bright beam on its underside lit up the corner of one of the nuke storage buildings. This illumination lasted several seconds before the beam disappeared. Then the object moved on toward the base's logistics yard, railway station and cement factory, but at some point, it returned to the weapons depot and hovered over it again, this time a little higher, before it suddenly accelerated and flew off.

In all, the witnesses had the UFO in sight for nearly two hours.

The Rockefeller Ending

This incident, actually told in brief here, was so detailed and had so many witnesses that it attracted the attention of a very unusual UFO enthusiast: Laurance S. Rockefeller.

The brother of onetime U.S. vice president Nelson Rockefeller and a member of one of the most powerful families in American history, Laurance Rockefeller paid for the aforementioned report: "Unidentified Flying Objects Briefing Document: The Best Available Evidence" in 1993. Authored by Don Berliner, Mary Galbraith and J. Antonio Huneeus, it looked into more than twenty cases that, in the opinion of the authors, contained the most reliable evidence for the paranormal nature of UFOs.

Because of his political connections and undeniable influence, Rockefeller was able to present this study to President Bill Clinton in 1995, along with a request to have the government release any pertinent files it had regarding the existence of UFOs.

But what Clinton did with the Rockefeller-funded document, if anything, remains unknown.

16

The Secret at
Huangyangtan

The Mysterious Photo

One day in 2006, a German researcher was looking
through Google Earth images of China when he came
upon something very puzzling.

It was a high overhead shot of a large piece of terrain
located in Huangyangtan, a sparsely populated region
of China about 550 miles west of Beijing. The image
showed some odd shapes on the ground.

These shapes looked like typical mountains, valleys
and rivers—but something was not right. The propor-
tions seemed off.

It took a while before the researcher realized he was

not looking at a natural landscape but rather a large, extremely elaborate, highly detailed scale model of a piece of rough mountainous terrain.

China has always been a land of mystery, even more so today with its repressive yet enterprising government immersed in seemingly endless rounds of political secrecy and military intrigue.

But what was this?

Subsequent photos showed vehicles and support buildings close to the sophisticated scale replica; only then was it clear the model had been built on the site of a previously unknown military complex located at Huangyangtan.

Further study determined the model was an exact one-to-five-hundred-scale re-creation of the Aksai Chin region, a forbidding piece of mountainous territory along the Chinese-Indian border. The facsimile, which represented an area measuring about three hundred by two hundred miles (or roughly the size of Switzerland), featured precise duplicates of snowcapped Himalayan Mountains, cavernous valleys, icy tributaries and the many frozen lakes that make up the Aksai Chin region.

But identifying what the mock-up depicted only deepened the mystery.

Aksai Chin is a long way from Huangyangtan, more than 1,500 miles to the west to be exact. The region is virtually uninhabited. There are just a few villages and small settlements there and, because it's squeezed between the Himalayas and the Karakoram Mountains, the weather is pretty much horrible all the time.

In fact, up to this point Aksai Chin had been known for just two things: It's one of the most treacherous and isolated places in the world, and back in 1962 China and India fought a war there.

Battle Across the Mountaintops

The Sino-Indian War started on October 20, 1962, but even after fifty years exactly what caused the conflict is hard to determine.

There had been a few minor shooting incidents in the late 1950s between China and India along their common frozen border. Plus, India had recently granted asylum to Tibet's religious leader, the Dalai Lama; this after the Chinese government had taken over Tibet.

But these events don't seem to be enough to start a war. Yet start a war they did.

Chinese troops invaded India at several points along the border, including the Aksai Chin region. While the

Red Army overran the defending Indian forces fairly easily, much of the fighting was done under almost unbelievably harsh conditions. Some of the combat took place at altitudes of more than fourteen thousand feet. No surprise, both the Chinese and the Indians had more troops die due to the freezing weather than due to each other's bullets.

The war lasted just a month. Then the Chinese simply declared victory and immediately withdrew to their side of the border. Soon afterward, U.S. President John F. Kennedy proclaimed that the United States would defend India if the Chinese attacked again. And about a year later, the new U.S. president, Lyndon Johnson, talked about giving nuclear weapons technology to the Indians to help them thwart any repeat attack by the Chinese. But that was about the end of it.

These days, the odd little war around Aksai Chin seems long forgotten. No shots have been fired there in a long time and both countries have only the barest military presence in the area.

Empty Theories

So why would the Chinese create such a scrupulously accurate facsimile of this godforsaken place?

One theory was that the Chinese military built the mock-up to train its tank drivers. But the scale of the model is too diminutive for tanks to drive through. Plus, again, the Aksai Chin region features some of the most frightening terrain on Earth. High mountains, deep gorges, few roads, lots of snow and ice. Not exactly tank country.

Another theory said the model was intended to train Chinese guerrillas who would someday be dispatched to the border to destabilize it. But knowing the area is frequently under assault from terrible blizzard-like weather, it's not ideal guerrilla country either. Plus, who would be sent there to attack?

The third theory is that the Chinese military constructed the model as a training aid for its helicopter and fighter pilots. But no aircraft were used during the 1962 war, mostly because of those inhospitable conditions, so what would change now? Besides, why would China choose to fight another war in this extremely remote place? Both China and India have nuclear weapons these days; both have nuclear missiles capable of reaching the other's capital.

If they were to go to war again, chances of it being a conventional war, fought among the Himalayan mountaintops, are all but nil.

The $100 Solution

In a previously published interview that appeared in the *Sydney Morning Herald* in July of 2006, Michael Barlow, deputy director of the Defense and Security Applications Research Center, said that although models of treacherous terrain were used for military training a long time ago, he'd never heard of anyone building one in modern times.

"The only large-scale models built these days are of urban environments," he observed. "Plus, you can literally buy a computer game for a hundred dollars and load in any existing terrain for the purposes of military simulations. Any good computer program could re-create the same thing that exists at Huangyangtan."

So if this model was not built to train tank crews, guerrillas or pilots, why was it built?

No one seems to know—except the Chinese themselves. And so far, they're not talking.

But maybe the real question is, why is the Aksai Chin region suddenly so important to the Chinese, again, if not for standard military purposes?

The answer might not be easily found in Beijing, but that's not the case in those remote villages up in the

Aksai Chin's mountains, the place the Huangyangtan model was built to re-create.

In fact, the locals are very forthcoming when asked why their remote piece of the world is so special.

They have an interesting story to tell.

The *India Daily* Article

There is an article that can be found on many UFO-themed websites that is said to be from the newspaper *India Daily*. While a few have questioned its accuracy, it nevertheless gives a fascinating account about what might be going on around Aksai Chin.

In fact, its headline says it all: **China and India both know about underground UFO base.**

The article talks about a place called Kongka La, a low-ridge pass within the disputed area of the India-China border. While the northeastern part, Aksai Chin, is under Chinese control, and India controls the south-western part, known as Ladakh, neither country has a lot of troops watching the border region.

But the article also says this area is well known to the locals for lots of UFO activity. Specifically, people frequently see UFOs around Kongka La, and local guides regularly tell tourists that strange well-lit triangular craft are seen flying out from underground locations near there all the time.

The article goes on to claim that the local people are amused when asked about this UFO activity and are surprised that both India and China try to hide the obvious. The locals say the extraterrestrial presence is well known, is deep underground but that both the Indian and Chinese governments want to keep it secret. Whenever the locals bring up the subject to government officials, they are told to keep quiet.

As fantastic as the article may seem, it's not the only media that has reported strange activity in the area over the years. For instance, in 2002, the Associated Press reported from the nearby Indian state of Uttar Pradesh that a flying sphere emitting red and blue lights repeatedly buzzed villages in that poor region, even attacking residents on occasion.

Centuries of Sightings

Good or bad, malevolent or benign, strange events happening in this Himalayan region are nothing new.

The nearby Nepalese have paintings going back a

thousand years depicting demons living inside the local mountains. The mysterious discs of Baian-Kra-Ula were found fairly close by on the Chinese-Tibet border.

And in the year 1661 a Jesuit missionary from Belgium named Father Albert d'Orville was visiting a Buddhist monastery in the area when he witnessed an astonishing event.

"My attention was drawn to something moving in the heavens," he wrote in a letter to a friend more than 350 years ago. "At first I thought it might be a species of bird unknown to me that lived in these regions. Then the object came near, taking on the form of a Chinese double hat and all the while rotating silently as it was being conveyed through the air on invisible wings. This visitation was definitely a thing of wonder or a trick. The object winged its way above the city exactly as if it wished to be admired. It circled twice and then was suddenly shrouded in fog and as much as I strained my eyes I could no longer see it."

The priest then asked his guide about what he'd just seen.

The guide replied, "My son, what you witnessed just now was not magic, because beings from other worlds travel across the oceans of space . . . and often come to earth near our monasteries."

What the Children Know

Possibly the most telling piece of evidence that something very unusual happens on a regular basis in the Aksai Chin region comes from another news report from the area, relayed by many media sources, concerning a drawing contest for young children held at a local school.

Told one day to draw things from everyday life, more than half the students drew strange objects in the sky or coming out of the mountains.

According to the locals, the children not only draw these things all the time, but many of them know what to look for and even where and when to look.

Speaking on his website, UFO author and researcher David Icke confirms such things while telling of his forays into the area.

He, too, had heard of the drawings being made by the local children and that the illustrations depicted craft that looked like domed UFOs.

But he points out that these drawings are very important because they came from kids who are pretty much isolated from the rest of the world—and in this case, that's a good thing.

"Their villages were very simple," Icke says on his

website of his visit to Aksai Chin. "We were few days' walk away from tourist resorts where TV and things of that nature were common. These kids were not watching American TV and had no books or comics to get references from."

The Colorado Connection

Why would there be so many UFO reports in this remote part of the world?

Surprisingly (or maybe not), we go back to San Luis Valley researcher Christopher O'Brien's words about what happens over those areas of the Earth where tectonic plates come in contact with each other. As O'Brien told us previously, unusual tectonic conditions seem to go hand in hand with lots of UFO sightings—and according to a United States Geological Survey website (www.pubs.usgs.gov): the Himalayas feature some of the most dramatic plate-tectonic forces in the world, with the Eurasian plate and the Indian plate continuously converging on one another.

These conditions create what some UFO researchers believe are open areas deep underground where UFOs could hide and operate from.

Back to the Mystery Model

Recent reports from the Aksai Chin area say more and more Indian military personnel on the ground are coming forward to say they're now seeing UFOs on a daily basis.

Indian helicopter pilots have been complaining about sophisticated electronic jamming going on in the area. Some Indian Air Force pilots are also reporting seeing strange flying objects above the Himalayas around the Chinese border. Also, YouTube is filled with videos of alleged UFOs flying over the entire Himalayan region.

So, there's little doubt something highly unusual has been happening for some time in the Aksai Chin region.

Which brings us back to the scale replica of Aksai Chin at Huangyangtan.

Again speaking on his website, David Icke asks the question that's on everyone's mind: "I am intrigued as to . . . why the Chinese would map the area so accurately in a huge reproduction scale model," he said. "Although the model at Huangyangtan is huge, it would not be practical for troop movement studies in this day and age with GPS and satellites. Why then would they build this thing?"

That's what makes this all so puzzling. We don't have a clue—at least not yet.

So, we're only left to consider this: Is it odd that this insanely isolated area, this place where a very strange war was fought, where Earth's crust is probably thicker than anywhere else on the planet, where kids draw UFOs as part of their everyday life and where people have been seeing UFOs for hundreds of years, is the same place the Chinese chose to build an elaborate model of?

Yes, it is odd.

Make that *very* odd.

17

UFOs over Oz

The Haunted Highway

On January 20, 1988, Fay Knowles and her three sons set out from the city of Perth in Western Australia, heading for Melbourne in the east.

Their mode of transportation was a 1984 Ford sedan. Their route took them across the Nullarbor Plain.

Connecting South and Western Australia, the Nullarbor Plain runs for more than one thousand miles east to west. It's mostly flat, arid desert, and as its Latin name suggests, it's practically treeless. The Eyre Highway, which features the longest stretch of road in the world (ninety miles without a bend or curve), is its main highway. But despite the excellent condition of this and other roadways, crossing the Nullarbor Plain can be hazardous. Run out of gas or have a mechanical problem and your car will likely join the many abandoned vehicles that can be found along the way. At the very least you'll have a long walk ahead of you.

But the Nullarbor Plain is known for something even more unpleasant than breakdowns in the middle of nowhere.

Some of the most frightening UFO incidents ever reported have happened here—as the Knowles family was about to find out.

Driving through the Nullarbor at night, when it was cooler, the Knowles had proceeded on their trip incident-free.

But just before four in the morning, the family saw a bright light on the road ahead of them. Sean Knowles, twenty-one, was driving. They were somewhere between the towns of Madura and Mundrabilla.

Everyone in the car assumed the light was just a truck coming in the opposite direction. But then the light began moving erratically. It would disappear only to reappear an instant later, much closer to the Knowles's car. After this happened a few times, Sean floored the accelerator. The light disappeared—only to wind up behind them.

Sean hit the gas again, this time to get away. But an instant later, the bright light was in front of them again.

The family finally realized this thing was a solid flying object—and it was now blinding them with its intense bright light. Still, Sean was able to turn the car 180 degrees, intent on speeding away in the opposite direction.

But the object stayed with them. Sean performed another U-turn, nearly colliding with a car coming in the opposite direction, but again the object persisted in chasing them.

Then the family heard a loud thud on the roof of their car. Something had landed there! Though panic-stricken, Mrs. Knowles rolled down her window and reached up to the roof, only to feel something warm and spongy. When she pulled her hand back inside, it was covered with hot black soot. An instant later, the car was filled with a black mist that stunk of rotting flesh, accompanied by a painful high-pitched sound. Most bizarre of all, the family's voices became lower in pitch and their speech started to slow down.

With all four members beyond terrified, the object began *lifting* their car off the roadway. It carried the Ford for a short distance before slamming it back down to the pavement with such force that one of its tires blew out.

Somehow, Sean managed to pull to the side of the road, where the family scrambled out and hid in some bushes. They watched as the object hovered around their disabled car for a few moments before taking off at tremendous speed. After staying hidden for another fifteen minutes, the family summoned the courage to put on their spare tire and flee to the nearby town of Mundrabilla.

Luckily for the Knowles family, they met three witnesses who'd also seen the strange object out on Eyre Highway. Returning to the scene of the incident, one witness saw skid marks and located the Ford's blown-out tire. When police questioned the family members, they found them severely traumatized. They were convinced something truly bizarre had happened to them.

But theirs was not an isolated case. So many cars have been chased by flying objects along the Nullarbor Plain that one local government erected a sign warning motorists to "Beware of the UFOs."

But why here? What is it about the Nullarbor Plain that attracts this sort of nasty UFO?

Someone's Watching Woomera

Not far from the eastern end of the Nullarbor is the Woomera Test Range.

Though its name has changed several times over the years, it was here in the 1950s that the British military, in dire need of wide-open spaces, came to test its newly developed atomic weapons.

It was a controversial decision. Not only would parts of Woomera be inundated with radioactive fallout as a result of the tests, but the site itself also had been previously inhabited by Australian Aborigines, all of whom had to be relocated against their wishes. This led to many disputes with the Australian government.

But one thing was not in dispute: As soon as the British nuclear blasts commenced, UFOs appeared over Woomera and the Nullarbor Plain in disturbing numbers—and they've maintained a presence there ever since.

These strange flying objects did more than just suddenly materialize above the Australian Outback, though. No sooner did the British begin their bomb trials than the UFOs began interfering with the nuclear testing itself.

Carefully planned detonations had to be postponed,

sometimes for days, because electricity throughout the Woomera test site would suddenly go dead whenever the UFOs appeared overhead. (The nearby town of Woomera, for which the site was named, often found itself blacked out for hours as well, again due to UFO activity.) Similar blackouts happened over American ICBM bases in the 1960s and 1970s as well as in the former Soviet Union during the 1980s. Again, for reasons no one can explain, UFOs seem very interested in mankind's nuclear weapons capability.

Because the Woomera nuke tests were recorded by movie cameras, reports say UFOs haunting the test range were caught many times on film. On one occasion, more than fifteen thousand feet of film was said to have been taken of UFOs crisscrossing the test area. But while the film was promptly sent to Washington, DC, for analysis, what happened to it after that is unknown.

UFO activity over the area ebbed and flowed over the next few decades, but it never went away completely. Then another top-secret facility was built at Woomera. Called Joint Defense Facility Nurrungar, it was operated by both Australia and the United States not for testing nukes but as a ground station for space-based surveillance platforms. In other words, it controlled spy satellites.

But UFOs soon showed a high interest in this facility as well. In fact, from the time operations began at Nurrungar in 1969, UFO activity in the area skyrocketed back to the same intensity as the days of British nuclear testing of the fifties.

When the Nurrungar base closed in 1999, many of its operations were moved 350 miles to the north, to another classified facility called Pine Gap. This place had seen its share of UFO incidents, too—but like a lot of UFO stories from Oz, these were anything but typical.

Weirdness at Dead Center

Pine Gap is considered one of the most secret places in the world.

Located in a small valley at the foot of the Mac-Donnell Ranges, the facility (officially named Joint Defense Facility Pine Gap) is about twelve miles east of Alice Springs, a small town considered the geographic dead center of Australia. Like a lot of noncoastal Down Under, the surrounding area is hot, desolate and inhospitable.

Originally opened in 1970, Pine Gap has grown steadily over the years. Most often described as a satellite control and processing station, the facility these days is basically a larger, more powerful version of Nurrungar. Like that shuttered facility, Pine Gap is operated jointly by Australia and the United States, and, frankly, it's from here that America's newest fleet of spy satellites is run. And many of these satellites are designed to eavesdrop not just on other nations but also on targeted individuals as well.

Pine Gap consists of a large complex that includes up to eighteen giant igloo-like radomes, each hiding an extremely powerful antenna. About a thousand employees of the National Security Agency and the CIA work here, and what they do is considered so secret that the airspace around the facility has been declared a no-fly zone—the only such flight restriction in all of Australia.

A massive security fence surrounds Pine Gap, and its perimeter is patrolled both by American and Australian guards. Meanwhile, security inside is said to be so severe that some have claimed personnel are actually hypnotized to ensure they will not talk about what goes on there.

This last point is interesting, because there are many rumors that more things go on at Pine Gap than just controlling spy satellites.

Death Ray Hotel?

One frequent claim about the Pine Gap base is that a borehole some five miles deep has been drilled below the facility to house a massive underground antenna. This giant antenna might be used for emitting very low-frequency radio signals to the U.S. military or for some kind of mysterious subsurface research or, as some people assert, to tune a gigantic "wave field" around our planet.

Other rumored deep-secret projects at Pine Gap include ongoing high-energy, high-voltage research; construction of a "death ray" plasma cannon; electro-magnetic propulsion tests and even specialized "power broadcasts" that are able to fuel U.S. Navy submarines thousands of miles away. A large nuclear reactor is also alleged to be in place at the facility, hidden deep within an underground chamber.

Even stranger, some locals say enormous amounts of food arrive at Pine Gap by airplane from the United States on a weekly basis, while other eyewitnesses insist they've seen air transports unloading things like furniture, high-end appliances and other provisions that one would ordinarily expect to see in a plush hotel.

This has led to speculation that an entire city exists somewhere beneath the facility.

Whose UFOs Are They?

It's no great surprise, then, that Pine Gap has been the site of many UFO incidents, just as with Nurrungar and Woomera before it.

But here's the twist. Few if any of these stories depict unidentified objects buzzing the facility or interfering with its functions. Just the opposite. Many of these reports come from people claiming to have seen UFOs *operating out of* Pine Gap.

Some of the reports include the following:

> In 1975, passengers on a plane passing just outside Pine Gap's restricted airspace say they saw a large white object take off from a hidden tunnel on the base and fly off to the northwest.
>
> In 1980, two members of the Northern Territory police, who were taking part in a search for a missing Alice Springs child, saw a trio of bathtub-shaped objects appear over Pine Gap. The strange aircraft then disappeared, one at a time, into an opening in a hillside deep within the base.
>
> In 1989, three hunters were camped out for the night near Pine Gap. Just before dawn, the trio saw a large camouflaged door hidden in the side of a hill inside

the base suddenly open. A large disc flew out of this opening and climbed into the sky at fantastic speed. Once the flying machine disappeared, the door quickly closed behind it.

But if these are UFOs, then whose UFOs are they?

More than a few locals say they've observed discs about thirty feet in diameter being unloaded from large U.S. cargo planes at the airport serving Pine Gap.

So the question is, are UFOs somehow assembled at Pine Gap and flown from there? Could they be spy craft? Drones made to look like UFOs?

Like many things having to do with secret bases, it's impossible to tell.

But whatever the case, flying objects of uncertain origin are just the tip of the iceberg when it comes to strange things happening around Pine Gap.

Close Encounters of the Aussie Kind

According to a number of media sources, something *really* spectacular happened at the Pine Gap facility in 1984.

The story comes from five UFO enthusiasts who were reportedly tipped off by someone in the know that, on a certain night, a major event was going to take place at the base. With this in mind, the group was able to get within several miles of Pine Gap, and indeed, they claimed to have witnessed something literally out of this world.

Using binoculars, they first saw a number of people inside the base who were wearing coveralls and assembling out in the open.

Suddenly an intense beam of gold light appeared from the center of the base. This beam was at least fifteen feet wide and shot straight up into the night sky. Clouds began forming around this beam, staying in place despite a steady wind blowing in the area.

The beam started pulsing, causing the clouds to take on the appearance of massive smoke rings. Oddly, on seeing this, the witnesses reported, they all became extremely nauseated.

Then the witnesses saw five objects approaching the base from the south, flying at about one thousand feet. Four of these objects were described by the witnesses as star-shaped and arranged in a diamond formation. The fifth object, which was shaped like a cylinder, was following about two miles behind.

Once over the facility, the four star-shaped objects took up positions at each corner of the base. Then the cylindrical object arrived, lowering itself to about five hundred feet off the ground. Light beams began flashing all over the base and between the five objects, a display that lasted several minutes.

Then one of the small star-shaped objects landed at the northern end of the base. It remained there for nearly twenty minutes before taking off and resuming its original position.

Now the intense beam of gold light reappeared and one of the small star-shaped objects began to orbit it, as if examining it. As this was going on, the strange clouds returned and surrounded the gold beam once again.

After a few minutes of this, the gold beam disappeared for good and the weird smoke rings blew away. The cylinder-shaped vehicle rose in altitude as the star-shaped objects went back to their original diamond-shaped formation.

Then as a group, the five objects roared off to the south, vanishing in a matter of seconds.

CE3—The Sequel

Exactly what happened at Pine Gap that night in 1984? For those familiar with the 1977 Steven Spielberg movie *Close Encounters of the Third Kind*, it seems the only thing missing from the witnesses' account was composer John Williams's five-tone motif that has become almost as famous as the movie itself.

But . . . strangely enough, a strikingly similar event was reported at Pine Gap in 1973, eleven years earlier.

In this incident, a man camping a few miles from the base saw a vertical shaft of bright blue light suddenly appear from the middle of the facility. Grabbing his binoculars, he spotted a strange-looking object hovering about a thousand feet above the base. Circular and gray with a dome on top, this object was enormous in size—more than five hundred feet in diameter by the witness's estimation.

For the next half hour, the witness watched as a series of blue and gold beams bounced back and forth between the gigantic saucer and the base. Then the object suddenly began oscillating, lighting up like a neon sign.

An instant later, it shot straight up and was lost among the stars.

The UFOs Next Door

UFOs have also been frequently reported around Alice Springs, the small town about a dozen miles away from Pine Gap.

In 1998, people there watched a weird silver glow hover over their town for nearly a half hour before it roared off to the south. Also spotted over the town on several occasions have been triangular-shaped clusters containing five lights that flicker brightly before eventually fading away.

In 1996, up to *forty* orange lights comprising two distinct formations flew over the town. Half the lights formed a circle that was visible for about fifteen minutes. Then the remaining lights moved in from east to west, forming an evenly spaced straight line. Both formations then traveled west and soon went out of sight.

That same year, in an incident similar to that endured by the Knowles family, a UFO touched down about five miles west of Alice Springs. Witnesses driving nearby said the UFO was about seventy-five feet in diameter and colored blue with a flat base.

As the witnesses watched, a halo formed around the blue object. This was followed by a whirling noise. Suddenly the UFO took off, and an instant later it was hovering above the witnesses' car.

It stayed there for a few terrifying moments before finally flying off toward Pine Gap.

Read All About It

One particular UFO story from the Pine Gap area actually made front-page headlines Down Under.

It happened in 1976 when a number of media sources reported that a UFO had crashed near the Pine Gap base, killing all on board. The occupants' bodies were recovered, but as the story goes, they were not human. Witnesses claim the remains were immediately transported to Pine Gap.

As strange as this story sounds, it was widely reported by the Australian national media. Even the government's own TV channel promised further details on the incident.

But in reality, the government never mentioned the crash again, at least not officially.

The "Lost" Files

Exactly what Pine Gap's strange connection is to UFOs remains a mystery. Perhaps with the exception of Groom Lake itself, no other secret military base has been so directly linked with the operation of UFOs as this highly unusual place in the middle of the desolate Outback.

But making the whole UFOs Down Under story even more puzzling, in 2010, Australia's Department of Defence announced it had lost many of its records on UFO sightings, not just over Pine Gap but over the *entire* continent, files it had collected over several decades.

This episode began when a Sydney newspaper asked the Australian military to turn over the UFO files via a freedom of information request.

But after two months of searching, the Australian DOD admitted the documents could not be found and that some of the files had likely been destroyed—which according to the Australian military was part of normal administrative procedure.

Blame It on the Brits?

Australia seems to attract a different breed of UFOs than the rest of the world.

Positively demonic when it comes to incidents on the Nullarbor Plain, aggressive and proactive when shutting down the Woomera weapons tests and yet oddly cooperative and even collaborative over Pine Gap—at least according to witnesses who say they've seen strange things happening there.

For want of a better description, UFOs over Oz seem a bit schizo. Friendly one moment, scaring the bejesus out of citizens the next.

Was it the British atomic tests of the fifties that opened this can of worms for the Aussies?

"There were certainly a number of civilian encounters with UFOs in Australia in the early 1950s and even before," author Nick Redfern told us in an interview. "But it's correct to say that with the beginning of the British nuclear tests at Woomera, UFOs began showing up with great frequency over secret military sites all over the continent, and that never really stopped. They're still doing so today."

18

Last Stop:
Playing the HAARP

Fooling with Mother Nature

A phrase Mark Twain was fond of repeating went as follows: "Everyone talks about the weather, but nobody does anything about it."

If you believe what some people are saying about a little-known U.S. government research project, though, Twain might finally get his wish—but not in any way he could have imagined.

The project is the High Frequency Active Auroral Research Program, or HAARP. It's located in an isolated corner of Alaska called Gakona, about two hundred miles northeast of Anchorage. Built on a tract of land once occupied by a massive U.S. Air Force radar installation, HAARP combines 360 radio transmitters with 180 antennas (each one sixty-eight feet tall), for what looks like a bizarre metal forest covering more than forty acres of remote bush country.

What does HAARP do? It depends on whom you ask.

The project's official website states, "HAARP is a scientific endeavor aimed at studying the properties and behavior of the ionosphere, with particular emphasis on . . . enhancing communications and surveillance systems for both civilian and defense purposes."

In effect, HAARP directs huge amounts of energy at the ionosphere—the electrically charged sphere surrounding our planet, about fifty miles high—and then bounces that energy back to earth.

However, critics of the project claim that when that energy comes back down, it can penetrate the Earth's surface for many miles, revealing things like underground munitions, hidden tunnels and possibly untapped mineral caches—very little to do with enhancing communications.

Critics also cite that the radio frequency needed for this earth-penetrating magic is close to the same frequency that disrupts human mental functions. This frequency might also affect the migration patterns of certain fish and animals.

But most disturbing of all is the charge that what HAARP *really* does is heat up the ionosphere with tremendous amounts of energy in an attempt to control the weather and use this unnatural power as a bizarre and dangerous weapon.

"Owning the Weather"

Though the idea of using the weather as a weapon might sound like science fiction, it's not. The U.S. military has been studying the concept since at least the mid-1990s, as was made clear by Nick Redfern in his book *Keep Out*.

Redfern quoted from a paper called the "USAF 2025 Report." It was prepared in 1996 by the College of Aerospace Doctrine, Research, and Education Center at Maxwell Air Force Base in Alabama.

Said Redfern: "One particularly intriguing subsection of the report had the notable title of 'Weather as a Force Multiplier, Owning the Weather in 2025.' It states that the U.S. military has been hard at work trying to determine if the manipulation and even the creation of harsh weather conditions such as hurricanes or earthquakes, volcanoes and other forms of devastation might be considered as a viable tool of warfare in the very near future."

The History Channel was also on the story. A 2006 TV special titled *The Invisible Machine* detailed how heating up the ionosphere could indeed turn weather into a weapon of war.

"Imagine using a flood to destroy a city or tornadoes to decimate an approaching army in the desert," the documentary asked. "The U.S. military has spent a huge amount of time on weather modification as a concept for battle environments."

So the weather could be used as a weapon. But can HAARP really control the weather?

Critics say the facility has the capability to heat up an enormous part of the atmosphere—maybe four hundred square miles or more—to a staggering temperature of fifty thousand degrees Fahrenheit. Moreover, the way that HAARP is set up (basically as a grid of 180 transmission towers in what is called a phased array), this heat and energy can be directed at many points of the sky. In other words, if the critics can be believed, HAARP's atmospheric baking process can be *targeted*. And that process could involve, for instance, rerouting high-pressure systems or even adjusting the flow of jet streams, with catastrophic weather being the result.

So, if some unfriendly country is doing something the U.S. government disapproves of, is a massive hurricane heading its way? Or a devastating flood? Or even an earthquake?

Even E.T. Might Be Scared

What does this have to do with UFOs?

Maybe not much.

Like Tonopah, Nevada (that similarly mysterious place approximately 2,500 miles to the southeast), HAARP is not a common locale for UFO sightings—which is intriguing in itself.

True, some conspiracy enthusiasts claim that HAARP's real purpose is to create a ginormous electromagnetic shield around Earth to prevent an attack from outer space. And others swear that HAARP has *already* induced earthquakes around the world and that UFOs spotted over the targeted areas before they were struck hint at some kind of collaboration. But if anything close to what some people claim HAARP can do is true, then frankly, an attack from outer space might be preferable.

What is particularly frightening about all this is another fact that Nick Redfern dug up and wrote in his book, *Keep Out:* "On April 28, 1997, then U.S. Secretary of Defense William S. Cohen gave the keynote speech at a conference on 'Terrorism, Weapons of Mass Destruction and U.S. Strategy' held at the University of Georgia. Cohen must have shocked his audience by telling them, 'Powerful shadowy forces [are] out there engaging in an eco-type of terrorism whereby they can alter the climate and set off earthquakes and volcanoes remotely, through the use of electromagnetic waves.'"

Cohen added, "There are plenty of ingenious minds out there that are at work finding ways in which they can wreak terror upon other nations. It's real."

Wow . . .

That the United States is not the only entity working on this technology changes the dynamic dramatically. The European Incoherent Scatter Scientific Association operates an ionospheric heating facility near Tromso, Norway. And Russia runs the Sura Ionospheric Heating Facility in Vasilsursk, near Nizhniy Novgorod.

Norway is considered a country friendly to the West, and since the early 1990s Russia is no longer the boogeyman it once was.

So who was Secretary Cohen talking about when he said: "Powerful shadowy forces out there . . . can alter the climate"?

Open or Closed?

According to HAARP's management, the project strives for transparency. All of its activities are logged and publicly available. Scientists without security clearances, even foreign nationals, are routinely allowed on the HAARP site.

In addition, the HAARP facility regularly hosts open houses during which any civilian may tour the entire facility. Furthermore, scientific results obtained by HAARP are routinely published in major research and defense journals.

However . . . when former governor of Minnesota and present-day conspiracy theorist/TV star Jesse Ventura (who has openly questioned whether the government is using the Gakona site to manipulate the weather) made an official request to visit the research station, he was denied.

And when Ventura and his film crew showed up at HAARP anyway, they were turned away.

So much for transparency.

Too Dumb to Understand?

Various mainstream scientists have commented that HAARP makes an attractive target for conspiracy theorists because, in the words of one skeptic, "Its purpose seems deeply mysterious to the scientifically uninformed."

But maybe that's the whole idea. Maybe instead of miles of desert as a buffer, or rings of barbed wire, or a small army of private security guards, maybe this unsecret place in Alaska is the most dangerous place of all, and for cover it uses that elitist, look-down-one's-nose comment perfectly delivered in a Harvard accent: "Sorry, this is deeply mysterious to the . . . uninformed."

The Future and Weather Wars

But uninformed or not, a dangerous clock might be ticking. Take the example of the United States and the A-bomb. Once the first nuclear device, developed via the legendary Manhattan Project, was exploded in 1945, America instantly and indisputably became the most powerful nation on Earth—and automatically held an atomic monopoly over the rest of the world. Happy days indeed.

But then Russia exploded its own A-bomb in August 1949, and that monopoly came to an end after just four years. Other countries have since joined the nuclear club (China, Pakistan and North Korea among them), making America's claim of pure world dominance through nuclear weapons a thing of the distant past.

So what happens when China or North Korea or even Iran acquires the capability to use and control the weather?

Then what will we have? A war of competing HAARPs? Hostile countries hurling massive floods or hurricanes at each other? Or creating über-destructive droughts or earthquakes to soften up an opponent before an attack?

The weather is an extremely powerful thing. It can likely do as much damage, take more lives and affect the long-term environment and the Earth itself as can today's arsenal of nuclear weapons.

Is this really how we want to fight the wars of the future? By changing the weather?

That's probably not what Mark Twain had in mind.

19

The Second Meeting

Another Beer with the Spook

As agreed, my Spook friend and I set up a time and place to meet again.

He'd read a draft of this book as promised and forwarded some positive if general comments. He'd been especially interested in the sections on Russia and also provided new information pertaining to UFOs over Scotland.

I was glad to get his feedback even though it was broad in nature. I really didn't expect anything else. I knew he would not floor me with some incredible revelation like he and his colleagues knew what UFOs were, or where they came from, or why they seem to be attracted to secret bases around the world.

Still, in the time leading up to this second meeting, I couldn't stop thinking about all the bits of information, the stories, the interviews, the anecdotes and the plain old reading I'd done while putting this book together and concluding that many of the events I wrote about just seemed too bizarre. Too far-fetched to be real.

More than once, I fretted that the whole UFO mystery might be nothing more than a grand expansion of the legend of Ong's Hat. Maybe *that* is what's really happening. For whatever reason a person makes up a UFO story from whole cloth and tells it to another person, and then they add something and it goes on to another, who adds their own embellishments, and on and on, and pretty soon we have an elaborate tale about a UFO doing this fantastic thing, or that fantastic thing, and the whole story is so detailed that some people can't help but take it all as fact.

As Michael Kinsella reminded me about Ong's Hat, "Legends force audiences to choose how to interpret them . . . and therein the 'truth' or 'reality' of the story comes to life."

So, I found myself asking the same question over and over. Is that all the UFO puzzle is? Just one huge, elaborate game of telephone?

But then I thought, *No.* Not *all* these stories are made up. Not all can be cases of misidentification of aerial phenomenon, or delusions, or hoaxes, or people just trying to cash in because no one is looking over their shoulder to verify what they claim has been already verified.

I know from my first UFO book, *UFOs in Wartime*, that pilots see these strange flying things all the time. That sometimes military pilots see UFOs in the midst of combat. And airline pilots see them just as often while shepherding hundreds of people across the sky. These pilots are expertly trained professionals, trusted with the controls of multimillion-dollar aircraft. They *know* the difference between a strange-looking cloud, the planet Venus at sunset, the reflection of another aircraft twenty miles away and something they've never seen before.

When you read accounts like the famous 1956 "Gander Sighting" off Newfoundland, when a hundred Navy personnel, many of them pilots themselves, saw a huge saucer flying alongside their plane, you know they couldn't possibly have been imagining it. When hundreds of people living in the San Luis Valley routinely see UFOs (and a whole lot more), they all can't be making it up. Not everyone in Bonnybridge or Gorebridge, Scotland, is a kook or a little too much into the grain spirits. And what Russian soldier would lie to the KGB about seeing a UFO over a nuclear storage facility? Why would Marina Popovich put her estimable reputation on the line by speaking out about things that weren't true? Why would the kids in Aksai Chin draw pictures of strange things flying in the sky if they didn't see them so often?

No, for every story about Dulce Base, there's a story about hundreds of people seeing an unexplainable parade of UFOs stream across the Bahamian skies. For every guy like "O'Hennessy" and stories of J-Rod 52 there's a serious investigator like Christopher O'Brien, Nick Redfern, Jerome Clark or Bill Birnes and the dozens of UFO radio commentators, film producers and authors. People who have devoted their lives to trying to solve this mystery. They all can't be wrong. They all can't be just tilting at windmills. It's in the numbers. All it takes is for one UFO sighting to be true; then in effect, they're *all* true.

So yes, Mulder, something is out there. Something is happening. UFOs do exist; we just don't know what they are yet.

* * *

In anticipation of my follow-up meeting, I thought about the specifics of some of the places visited in this book. Secret places around the world that for the most part seem to have that one thing in common: an unexplainable connection to UFOs. If the most important question is "What are UFOs?" then maybe the second most important question in the big scheme of things is "Why are they attracted to these secret places?"

Again, it's in the numbers. This book is full of instances showing that it *does* happen a lot, that these things *are* attracted to secret places, but even after this journey we still don't know *why*. The entities that fly, control, manipulate these things we call UFOs—are they just looking in on us? Are they all-powerful yet benign observers? Or could they be something as simple as "time tourists"—that is, people from our own future who have come back to see history as it's being made? This is probably the most optimistic theory of all because, as a very learned friend once told me, at least it means we *have* a future.

Or could it be something more sinister? Are UFOs showing up over these secret places—just as they show up over ICBM bases, nuclear power plants, nuclear weapons factories and other high-priority military targets—because they are just doing what any good military commander would do: performing intensive and extensive reconnaissance before some kind of attack begins?

And if that's the case, how and when will we find out?

It Gets Weird, One Last Time

My second meeting with the Spook started just as the first one, with a couple beers in a half-empty bar on a weekday afternoon.

The location was not the same, though; it was nowhere near the waterfront bar where our first meeting had taken place.

This new location was a bit surprising, and at first glance as far removed from the topic of UFOs and secret bases as I could have imagined.

But there was a point to it, as I was about to find out.

We finished our beers and walked out of the bar, down the crowded sidewalk, across a busy street and through an iron gateway.

A minute later, we were sitting on two small park benches pulled together.

We were surrounded by redbrick buildings neatly

laid out between geometrically perfect patches of finely trimmed lawns and lots of shady trees. This was a busy place. Many people were walking about—men and women, but mostly young people in their early twenties. By design, we'd put ourselves out of earshot of all of them.

Once settled, my friend returned to a subject he'd briefly mentioned at the bar. Glancing around our new location and at all the people passing by, he said, "You'd be surprised how many here work for the agency."

I knew he meant the CIA.

"Or have worked for them," he continued. "Or how many come here for a year or two. Or some for even longer. There's a lot of them here, though. More than people realize."

As I was taking this in, because I was indeed surprised to hear it, my Spook friend told me this story.

Years before, closer to the beginning of his intelligence career, he was on assignment at an air base in Florida that must go unnamed. One day he passed by one of the base's buildings and noticed it was surrounded by concertina wire and heavily guarded by Air Force security personnel. The building, which was ostensibly a hospital, was also festooned with hazmat signs.

My friend had heard the rumors that something had

been hidden away at Homestead Air Force Base all these years. Maybe what Nixon had shown Jackie Gleason, or maybe something else. He'd also heard that whatever it was had been moved to this other Florida base after Hurricane Andrew basically wiped Homestead off the map in 1992.

So when he later got in a conversation with one of the security people assigned to this mysterious "hospital," he asked, half serious, "Is this where they brought the UFOs from Homestead?"

The security man just shook his head and replied, "If we had anything like that, we wouldn't hide it here. We'd hide it in plain sight."

And that was the point. Megiddo, all over again.

"Sure, there are secret bases," my Spook friend told me, now with a couple decades of intelligence work under his belt. "But when you think about it, a lot of people know where they are. Your book is a good example. Most of those locations are very well known. It's what's done inside them that's 'secret.'

"But just as that security guy told me, if something is *really* valuable or if something is *so* high priority it has to be hidden away at all costs, the best way to hide it is to hide it in plain sight. Someplace where no one would ever think of looking for it. Someplace where it couldn't be found by someone like you. *That's* the way these things work."

He looked around us and with a straight face said, "Maybe even in a place like this."

He pointed to the redbrick buildings and continued. "It could be hidden there, or there, or there. . . . You'd never know."

And for the last time, at least for this project, that's when it got really weird.

Because where we were sitting when he was telling me all this was in the middle of Cambridge, Massachusetts, in a place called the Old Yard . . . of Harvard University.

BIBLIOGRAPHY

The following books, articles, websites and interviews were extremely helpful to me while writing this book:

Clark, Jerome. *The UFO Encyclopedia: The Phenomenon from the Beginning.* 2nd ed. 2 vols. Detroit: Omnigraphics, 1998

O'Brien, Christopher. *Secrets of the Mysterious Valley: An Investigator's Journey Through the Unknown.* Kempton, IL: Adventures Unlimited, 2007

Redfern, Nick. *Keep Out! Top Secret Places Governments Don't Want You to Know About.* Pompton Plains, NJ: New Page Books, 2012

S4 and Area 51
"The Open Mind: The Aquarius Project" (2007) Exopolitics: How Does One Speak to a Ball of Light?
Interview with Norio Hayakawa, May 3, 2012
http://boulderexo.com/burish.html

BIBLIOGRAPHY

www.examiner.com/article/testimony-of-cia-assassin-recruited
    -from-navy-seals-goes-online-with-documents
www.examiner.com/article/cheney-taken-inside-s-4-to-view
    -flying-saucers-ebe-bodies
http://ufo.whipnet.org
www.unmuseum.org
www.ufohypotheses.com/s4informers.htm
www.bibliotecapleyades.net
http://ufos.about.com/od/ufofolkloremythlegend/p/russia1989
    .htm
www.ufoevolution.com
www.viewzone.com
www.freeworldfilmworks.com/fwa-area51.htm
www.illuminati-news.com/philip-schneider.htm
www.bibliotecapleyades.net/exopolitica/exopolitics_kennedy02
    .htm
www.nytimes.com/2009/07/14/us/14intel.html

San Luis Valley
O'Brien, Christopher. *Secrets of the Mysterious Valley.* Kempton, IL:
    Adventures Unlimited, 2007
Interview with Christopher O'Brien, May 11, 2012

Dulce Base
Bishop, Greg. "Dulce Base Was/Is Real." www.ufomystic.com
    /2010/09/24/dulce-base-was-is-real/, September 24, 2010
Bishop, Greg. "Official name for Dulce, New Mexico underground
    base   disclosed!!"www.ufomystic.com/2010/09/27/official
    -name-for-dulce-new-mexico-underground-base-disclosed/,
    September 27, 2010
Bishop, Greg. *Project Beta*. New York: Paraview-Pocket Books,
    2005
Bishop, Greg. "Report: Dulce Underground Base Conference."
    www.ufomystic.com/2009/03/29/report-dulce-under
    ground-base-conference/, March 29, 2009
"Dulce." http://aliens.monstrous.com/dulce.htm
"Dulce Underground Base." www.ufocasebook.com/dulce.html

"Dulce Underground Base." www.subversiveelement.com/Dulce
    Bishop2.html
www.Martiansgohome.com

Tonopah

www.mufon.com
www.nuforc.org
www.forteanswest.com
www.lazygranch.com
www.globalsecurity.org
www.dyn-intl.com
http://nsla.nevadaculture.org/index.php?option=com_content
    &task=view&id=704&Itemid=418
http://forteanswest.com/wordpress-mu/arizonalowfi/2011
    /11/16/tonopah-arizona-ufo-sighting-by-gasp-a-scientist/
www.nuforc.org/webreports/036/S36887.html
http://ufospottings.com/location/Tonopah-NV
www.nuforc.org/webreports/074/S74115.html
www.globalsecurity.org/military/facility/tonopah.htm
www.strangeusa.com

Homestead

Beckley, Timothy Green. "The Great One & UFOs." http://www
    .artgomperz.com/news2/greatone.htm, undated
Bower, Douglas. "UFOs-Aliens, President Nixon, and Jackie
    Gleason." http://voodoowhodo.blogspot.com/2007/12/ufo
    -aliens-president-nixon-and-jackie.html, December 20, 2007
Kennedy, William H. "The Occult World of Jackie Gleason."
    www.mysticvalleymedia.com/bonus.html, 2011
Knell, Bill. "Jackie Gleason's UFO Encounter." http://www
    .informantnews.com/modules.php?name=News&file=article
    &sid=453, May 2, 2012
Lobosco, David. "Jackie Gleason and UFOs." http://voodoo
    whodo.blogspot.com/2007/12/ufo-aliens-president-nixon
    -and-jackie.html, August 1, 2011
Murray, Marty. "Jackie Gleason's Trip to the Alien Morgue."
    http://rense.com/general70/gleason.htm, April 9, 2006

BIBLIOGRAPHY

www.phils.com.au/jackie.htm
www.presidentialUFO.com

Autec
Interview with Bill Birnes, May 8, 2012
www.greatdreams.com
www.motygido.co.uk/bahamas_lights.htm
www.bermuda-triangle.org/html/miami.html
www.ufospottings.com
www.disclose.tv/
www.mysteryoftheinquity.wordpress.com

Ong's Hat
Interview with Michael Kinsella, July 12, 2012
www.hijackedsignal.com
http://deoxy.org/inc2.htm
www.ufodigest.com
The Writings of Alexandra Bruce
www.deoxy.org

UK's Area 51
Interview with Nick Pope, May 15, 2012
http://news.stv.tv/tayside/181127-radio-ghost-mystery-at
    -former-raf-station
www.stone-circles.org.uk/stone
http://en.wikipedia.org/wiki/Crop_circle
www.paranormaldatabase.com/
http://teentweens.blogspot.com/2008/05/15-infamous-top
    -secret-bases-compounds.html
www.nationalufocenter.com
www.parascience.org.uk/investigations/hack/hack.htm
http://news.bbc.co.uk/dna/place-lancashire/plain/A593778
www.militaryairshows.co.uk
www.dreamlandresort.com/black_projects
www.nickpope.net
www.ltpa.co.uk/site_range/index.asp
www.secret-bases.co.uk/

Rosslyn Chapel

Interview with Andrew Hennessey, April 4, 2012

www.andrewhennessey.co.uk

www.ufoencounters.co.uk

www.ufoinfo.com/sightings/uk/090521a.shtml

www.unitedkingdomufogroup.co.uk

www.offtheplanet.blogspot.com

www.ufoexperiences.blogspot.com

http://www.caledonianmercury.com/

http://gerberink.hubpages.com/hub/Secrets-of-the-Knights
-Templar

www.europeupclose.com/article/the-secrets-of-rosslyn-chapel

Are UFOs Invading Scotland? http://dsc.discovery.com/

The Bell

Cook, Nick. *The Hunt for Zero Point.* London: Century Hutchin-
son, 2001

http://discaircraft.greyfalcon.us

http://sites.google.com/site/nazibelluncovered

www.abovetopsecret.com

www.fallenalien.com

Saddam's Area 51

www.bibliotecapleyades.net/exopolitica/esp_exopolitics_R_1_03
.htm

BBC. "1988: Thousands Die in Halabja Gas Attack." http://
news.bbc.co.uk/onthisday/hi/dates/stories/march/16/
newsid_4304000/4304853.stm, March 16, 2008

Calabresi, Massimo, and Timothy J. Burger. "Who Lost the WMD?"
http://www.time.com/time/magazine/article/0,9171,461
781,00.html, June 29, 2003

Gousseva, Maria. "Is Hussein Owner of a Crashed UFO?" http://
www.bibliotecapleyades.net/exopolitica/esp_exopolitics
_A_a.htm, January 31, 2003

NASA. "Space Shuttle Columbia and Her Crew." http://www
.nasa.gov/columbia/home/index.html, August 23, 2006

Norton-Taylor, Richard, and Julian Borger. "New theory for Iraq's missing WMD." http://www.guardian.co.uk/politics/2003/dec/24/uk.iraq, December 24, 2003

Trainor, Joseph. "Mysterious Lights Seen in Saddam's Area 51." http://ufoupdateslist.com/2003/mar/m27-003.shtml, March 26, 2003

Uncredited. "UFOs Fly as Allied Bombs Hit the Little Zab Valley." http://www.bibliotecapleyades.net/exopolitica/esp_exopolitics_R_1_04.htm, undated

Wagner, Stephen. "Saddam Hussein and the Paranormal." http://paranormal.about.com/cs/humanenigmas/a/aa041403.htm, undated

M-Triangle
Interview with Nikolay Subbotin
www.people.com
www.phantomsandmonsters.com
www.abovetopsecret.com
Yakimov, Valery. "My Experience of Observations of UFO and Anomalous Phenomenon in Perm Anomalous Zone." http://www.ufoevidence.org/documents/doc451.htm

Kapustin Yar
http://zighydevourer.wordpress.com/2011/10/12/1947-ufo-kapustin-yar-the-russian-area-51
http://ufos.about.com/od/ufofolkloremythlegend/p/russia1989.htm
www.ufoevidence.org
www.history.com/shows/ufo-hunters/videos/russian-roswell-inside-kapustin-yar#russian-roswell-inside-kapustin-yar
www.cosmostv.org/2011/08/russian-cosmonaut-marina-popovich-ets.html#ixzz20SD024rc

Huangyangtan
www.davidicke.com
www.indiaufo.blogspot.com
www.indiadaily.com/editorial

www.abovetopsecret.com
www.apinewsonline.com
www.smh.com.au

Pine Gap
Interview with Nick Redfern, May 30, 2012
www.strangedayz.co.uk/2007/09/ufos-on-nullarbor-plain.html
www.popularmechanics.com
www.standeyo.com
www.bibliotecapleyades.net
www.thewatcherfiles.com
www.telegraph.co.uk

HAARP
www.ufo-blogger.com
http://2012indyinfo.com/2012/01/19/strange-noises-are-they
    -ufos-thunder-haarp-or-a-hoax
www.forbiddenknowledgetv.com/videos/weapons/all-purpose
    -haarp-fake-ufosweather-war—missile-defense.html
www.haarp.net
www.wanttoknow.info/war/haarp_weather_modification_
    electromagnetic_warfare_weapons
http://beforeitsnews.com/alternative/2010/05/are-we-in-a
    -haarp-earthquake-war-20951.html
www.bariumblues.com/haarp1.htm